Ulrich Sidoine WUIBE WOUBASSI
Moïse ADAMOU
Elias NUKENINE NCHIWAN

Organic farming

Ulrich Sidoine WUIBE WOUBASSI
Moïse ADAMOU
Elias NUKENINE NCHIWAN

Organic farming

Insecticidal effects of three bioinsecticide powders on the entomofauna of eggplant in Bocklé (Garoua-Cameroon)

ScienciaScripts

Imprint

Any brand names and product names mentioned in this book are subject to trademark, brand or patent protection and are trademarks or registered trademarks of their respective holders. The use of brand names, product names, common names, trade names, product descriptions etc. even without a particular marking in this work is in no way to be construed to mean that such names may be regarded as unrestricted in respect of trademark and brand protection legislation and could thus be used by anyone.

Cover image: www.ingimage.com

This book is a translation from the original published under ISBN 978-620-3-44062-1.

Publisher:
Sciencia Scripts
is a trademark of
Dodo Books Indian Ocean Ltd. and OmniScriptum S.R.L Publishing group
Str. Armeneasca 28/1, office 1, Chisinau MD-2012, Republic of Moldova, Europe
Printed at: see last page
ISBN: 978-620-5-36627-1

DEDICATION

I dedicate this work to my beloved parents MAGNE Brigitte and WOUBASSI Gilbert.

FOREWORD

The present work took place in Bocklé in the North Region, Department of Benue, Arrondissement of Garoua 3e . I am grateful to Dr ADAMOU Moïse, Entomologist, Lecturer, Vice-Dean in charge of Research and Cooperation at the Faculty of Medicine and Biomedical Sciences of Garoua, for having accepted to direct this thesis, his availability, his advice and his contribution to the determination of insects. I am also grateful to Professor NUKENINE Elias NCHIWAN, Entomologist, Full Professor of Universities, Vice-Dean in charge of schooling and student follow-up at the University of Ngaoundéré, for his scientific rigour, the documentation made available to me and his contribution to the identification of insects.

My deepest gratitude also goes to :

- all the other teachers of the Department of Biological Sciences and all the administrative staff of the Faculty of Science, for their contribution to my training;

- my laboratory seniors (Mr. DÉLI KODJI Pra, Mr. YATAHAI Clément Mineo, Mrs. MASSAH DABOLE Odette, Mr. MOHAMADOU Moukhtar, Mr. YOUSSOUFA Ousmana, Mr. TCHOUBOU-SALE Abraham, Mrs. TCHOCGNIA TCHEPENI Fernande Cadette) for their collaboration;

- my fellow students (Miss ZENGMBE Jessica, Mr WARDA Ltanoua, Mrs SAH Charlotte, Mr YAGOUADA, Miss KEOU, Mr TCHOFFO) for their encouragement and collaboration;

- BEKOU & MAFOKWE families for their moral support and love;

- The LYBIPO family (TALONG Raoul, TAMDJOKOUEN Singor, ELANGUE Schilemann, OTU Ernest, MBANTANG Ameline) for their moral support and love;

- my brothers and sisters (Mrs TOUSSI Agathe, Miss MANEKOU Scole, Miss FOGANG Stéphie, Miss SOKOUDJOU Albertine & Mr. TAKAM Cédric) TAKAM Cédric); my friends (YOUAGA Marcelin, TAKAM Wilfried, FOKOU Roberto, FOMO Jacqueline, DIONGOU Dharlyne, MALLA Tatiana, SOUAIBOU Maestro, AMBRAOU Aboubakary, FIANGA Kimany, AVOMO Laetitia) for their assistance and moral support;

- my uncles (Mr FOTSING Antoine, Mr TENE), my aunts (DJOUHOU Marie, DJOUGANG Jacky, MANEKOU Sidonie) for their moral support;

- all my fellow Masters students for their contributions;

- all those who, from near or far, have contributed to the realisation of this work and whose names are not listed here.

SUMMARY

SUMMARY

In Bocklé (Garoua-Cameroon), from 01 August to 22 December 2020, the insecticidal effect of the powders of three plants (*Azadirachta indica, Carica papaya* and *Senna didymobotrya*) was observed on the entomofauna of *Solanum melongena in* order to assess the population dynamics of insect pests, to collect data on the number of leaves, flowers and fruits per plant, to study the foraging activity of the floriculture insect *Apis mellifera* and to determine the cumulative impact of botanical insecticides and this bee on the seed yields of this Solanaceae. The experimental field was arranged in a completely randomised block design with nine treatments repeated four times, namely: three aqueous extracts, one control and one synthetic insecticide (Optimal); four markings of 240 to 300 flowers of *S. melongena flowers*, two of which were differentiated by the presence or absence of protection of the flowers from insect visitation and the other two consisting of flowers protected and then opened exclusively to *A. mellifera on the* one hand and on the other hand protected from insects and then intended to be opened and closed without visitation by insects or any other organism. All the insecticides tested very significantly reduced the pest population and influenced the activity of flowering insects. *Azadirachta indica* caused a considerable decrease in the population of almost all bio-pests in the field but more so in Coleoptera and Lepidoptera, while *C. papaya* and *S. didymobotrya* reduced those of Hemiptera and Lepidoptera. These biological insecticides also contributed to the reduction of damage on leaves, flowers and fruits of this Solanaceae. A total of 15 insect species were recorded during our study period. 12 species of insect pests, of which *Arhopalus rusticus was* the most important with 24.25% of the number of individuals preferentially attacking the leaves and 3 species of insect pollinators recorded, of which *Apis mellifera was* the most important with 77.14% of visits. The activity of this bee extended from 8 am to 4 pm with two peaks of activity located between 10-11 am for the treated and control sub-plots and 12-13 pm for the sub-plots treated with *S. didymobotrya. Apis mellifera* collected pollen exclusively. The highest number of simultaneously active individuals was 1 per flower and 14 per 1000 flowers. The overall average visit time was 8.01 sec and high in the subplots treated with aqueous extracts of *C. papaya.* The overall average foraging rate was 11 flowers/min, low in both *C. papaya* treated and control subplots. For flowers left to pollinate freely and those protected from insects, there was an increase in the fruiting rate, the average number of seeds per fruit and the percentage of normal seeds, which varied respectively from 20 to 29%, 20 to 25% and 8 to 27% in the subplots treated with Optimal and *A. indica* due to the pollinating insects, especially *A. mellifera.* Through their cumulative action, powdery mildew insecticides and flowering insects increased the fruiting rate, the average number of seeds per fruit and the percentage of normal seeds which varied respectively from 10.86% for *A. indica* and 4.88% for the control; 50.59% for *C. papaya* and 20.03% for the control and finally 10.64% for *A. indica* and 1.20% for Optimal. Through its pollination efficiency, *A. mellifera* caused a significant increase in fruiting rates, average number of seeds per fruit and percentages of normal seeds respectively of 0% for the control and 28.58% for *C. papaya*; 55.29% for *A. indica* and 90.58% for *C. papaya* and finally 1.02% for Optimal and 40.16% for *A. indica.* In order to protect the aubergine crop from insect pests while preserving pollinating insects, the use of botanical insecticides should be encouraged as an alternative to the use of synthetic insecticides.

Key words: *Apis mellifera*, bio-insecticides, Bocklé, *Solanum melongena*, entomofauna, yield.

INTRODUCTION

The debate on the hierarchy of so-called essential occupations has been revived after the current health crisis (Puliighe & Lupia, 2020). Among them, organic farming has been put forward as a must in view of the need to feed the world (Puliighe & Lupia, 2020). The availability of food is one of our most basic needs (Puliighe & Lupia, 2020). To meet these needs, humans practice organic agriculture, including market gardening (Altieri & Nicholls, 2020).

Considered a food sovereignty activity (FAO, 2018), market gardening refers to farms growing vegetables and fruits for consumption (Equiterre, 2016) and is considered a component of urban and peri-urban agriculture in Africa (Muliele *et al.*, 2017). Depending on the edible part of the plant, vegetables can be grouped into leafy vegetables, fruiting vegetables and root vegetables. They are also presented as a highly specialised agriculture as they constitute one of the most productive agricultural systems in Africa (FAO, 2021) and have been booming for the past few decades in West Africa (Houadakpode, 2018). In Cameroon, it is a significant supplement to the diet, a source of income that complements income from the sale of cotton (IDEA, 2020), plays an important role in the fight against poverty and malnutrition, and contributes significantly to improving household income (Yolou *et al.*, 2015). Among these available vegetable crops, we have *Solanum melongena*.

It is a perennial plant that like all other plants is influenced by the severities of nature, notably abiotic or biotic factors in the environment (Yarou *et al.*, 2017) and African agriculture is under the threat of locusts and caterpillars (leconomie, 2021) which affect yields and the resulting post-harvest operations such as pest pressure (Mondédji, 2015; Daunay, 2016).

In order to make their production profitable, market garden producers use several methods to control pest attacks, among others chemical control (Houadakpode, 2018). Synthetic insecticides are the most used by producers because of their availability, ease of use and immediate effectiveness (Mondédji *et al.* , 2015). However, recent studies report that pre-harvest yield losses due to pests are estimated at 35% and will be about 70% in the absence of protection (Popp *et al.*, 2013). The uncontrolled use of synthetic insecticides produces undesirable effects on humans, the environment in general and the positive role that bees play goes unnoticed as do the decreasing populations of ladybirds, hoverflies and predatory spiders (INRA, 2019) and also Hymenoptera which are crucial for the maintenance of biodiversity in the world (Amoabeng *et al*, 2013) as about 80% of flowering plants are pollinated mainly by this order (FAO, 2020b). In view of this problem, it would be wise to develop alternative control methods that are inexpensive, effective and easy to adopt for Third World producers (Barry *et al.*, 2017) following the example of *Azadirachta indica* (Meliaceae), *Senna didymobotrya* (Fabaceae) and *Carica papaya* (Caricaceae), which appear to be effective in controlling pests while conserving pollinators (Barry *et al.*, 2017).

In Cameroon, for nearly three decades, the relationships between some floricultural plants and their insect pollinators have been increasingly known thanks to work carried out in the Adamaoua (Tchuenguem (2005); Adamou (2017); Djakbé *et al.* (2017; 2019); Zra *et al.* , 2020; Kingha *et al.*, 2021), Centre (Azo'o *et al.* , 2012a; Pando *et al.* , 2014; Douka *et al.* , 2017; Pharaon *et al*, 2021), Far North (Azo'o *et al.* , 2012b; Djonwangwé *et al.* , 2017 and Pando *et al.* , 2020; Azo'o *et al.* , 2021), Littoral (Taimanga *et al.* , 2018), North (Basga *et al.* , 2019; Kingha *et al*, 2021); North - West (Otiobo *et al.* , 2015), and West (Dongock *et al.* , 2011) on the one hand and on the other hand between plants and their insect pests in the field thanks to the work of Barry *et al.* , (2017; 2019) in the Adamaoua and Far North Regions.

Prior to our work, the floricultural entomofauna of *S. melongena* was studied in India by Kokopelli (2013); in France by Fayet (2017), Projet (2015); Paris (2016) where they showed that aubergine flowers are visited by solitary bees. Similarly, Charlotte *et al*, 2014; BSV (2019) in France; Esdras *et al*, 2015 in Côte d'Ivoire; ALENE *et al*, 2019 in yaoundé observed that aubergine is attacked by many pests including mites, thrips, bugs, nematodes and dyphophores.

Despite this work, information is lacking on the relationships between several insecticidal plants growing in Cameroon and their effects on pests and pollinators in the field.

The present work is therefore a contribution to the mastery of the relationships between botanical insecticides (*A. indica, S. didymobotrya* and *C. papaya*), insect pests and pollinators of *S. melongena* for their optimal management in Cameroon.

Specifically, it was about :
- inventory the entomofauna of *S. melongena* ;
- to evaluate the effects of botanical insecticides on the entomofauna of *S. melongena* in the field;
- to study the activity of *Apis mellifera* under the effect of aqueous extracts on the flowers of this species;
- to evaluate the influence of these extracts on the yield of aubergine;
- to assess the cumulative impact of botanical insecticides and floricultural insects on the yield of this plant;
- to estimate the pollination efficiency of *A. mellifera* under the effect of aqueous extracts.

CHAPTER I: REVIEW OF THE LITERATURE

I. 1 *Solanum melongena*

I. 1.1 General

Solanum melongena commonly known as aubergine is an annual herbaceous plant native to India belonging to the family Solanaceae (Joydeep, 2012). It is a dicotyledonous plant species cultivated for its edible fruits as vegetable or market garden plants that has been domesticated in Asia since prehistoric times 59 BC (Domi, 2018). It is a vegetable that can be grown in conditions of high temperature, sunshine, abundant light and water in well-drained soils enriched with organic matter and very humus (Binette & Jardin, 2019). The fruit is a berry that symbolises summer and can be of varying colour depending on the growing environment (IMFURA, 2019). The domestication of aubergine is probably ancient from wild populations of *S. incanum* Lam and *S. undatum* Lam, resulting in a very large diversity of varieties and cultivars but in any case *S. melongena*, does not exist in the wild. With a world production of 51,288,170 tonnes in 2016 (FAOSTAT, 2016), China (62.39%) leads followed by India (24.47%) and Egypt (02.32%) (OECD, 2019). The white variety is known as Dourga, Japonaise in the world, or Djagattou (in Bamiléké), Zong (in Eton, Ewondo, Bulu), Nga (in Bamoun), Hisingui (in Bassa) in Cameroon (Jangolo, 2017)

I. 1.2 Morphology

Aubergine is a perennial shrub reaching 150 to 200 cm in height. It is a strongly branched plant with a long taproot and leaves with or without spines that are densely covered with 8-10 star-like hairs (Daunay, 2010). Its leaves are alternate, simple; stipules absent; petiole 6-10 cm long; blade oval to oval-oblong, base rounded or cordate, often unequal, apex acute or obtuse, margin sinuously lobed, densely hairy (Daunay, 2012). Its stem is cylindrical, woody, cottony at the top, sometimes reddish purple, with or without spines (Joydeep, 2012). Its inflorescence is of the uniparous scorpioid cyme type with hermaphroditic sexuality (Mi-aime, 2020). The flowers are bisexual or functionally male, regular, white or purple, solitary and borne in the leaf axils. The fruits are berries of a wide variety of shapes (ovoid, pear-shaped, spherical, cylindrical and very elongated 2-35 cm long and 2-20 cm wide) and colours (from ivory-white, yellow, green, and more generally purple, purplish to almost black), uniform, graded or streaked. The floral formula of the studied aubergine is: 5S+5P+5E+2C (Jacques, 2021).

7

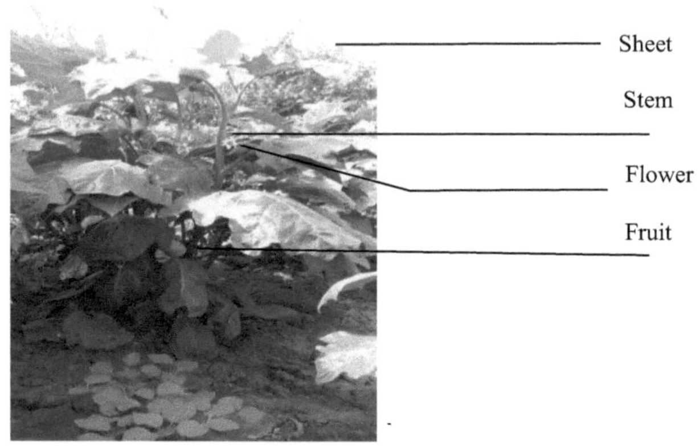

Figure 1: *Solanum melongena* plant (Bocklé, October 2020)

Figure 2: Flower of *Solanum melongena*

Figure 3: White spherical fruit

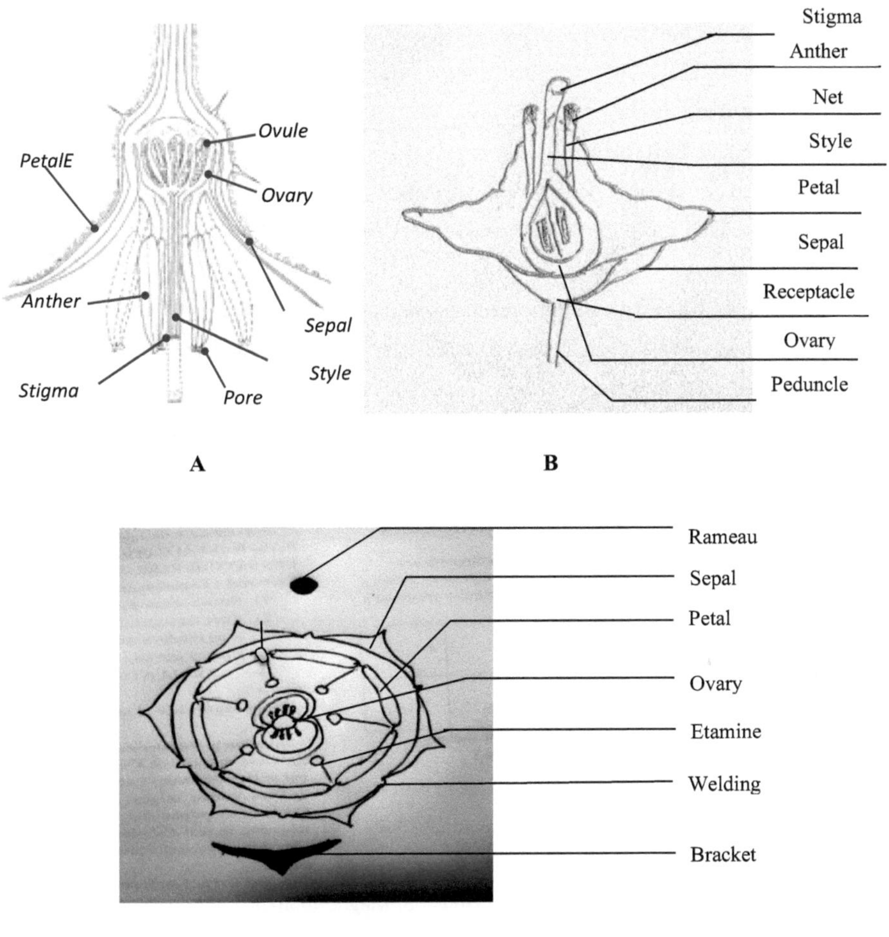

C (5S+5P+5E+2C)

Figure 4: Whole flower (A), Longitudinal section of the flower (B), Flower diagram (C) of *Solanum melongena*

I.1.3. Uses and importance of *Solanum melongena*

The aubergine is both a vegetable and a fruit that is of vital importance in our lives because of its multiple uses (OECD, 2019):

❖ **Economic importance**

It increases sources of income through marketing, with over 52.75 million tonnes sold worldwide. It is of paramount importance in Africa because of the job creation it offers to small farmers (knowing that yields can be very satisfactory even on a small area) which reduces the rate of stubble (FAO, 2021).

❖ **Nutritional use**

The aubergine fruit (berry) is used as a food for nutrition and can be eaten raw, grilled, fried or steamed as a stew with other vegetables. Nutritionally, its relatively balanced composition gives it an exceptional nutritional value (USDA, 2020).

❖ **Medicinal use**

Aubergine is also widely used for medicinal purposes. Different parts of the plant are used in decoction, powder or ash form to treat diseases such as diabetes, cholera, bronchitis, dysuria, otitis, toothache, skin infections, asthenia and haemorrhoids. It serves as a hypotensive, cholesterol regulator, laxative and anti-anemic. It helps in digestive diseases: dysentery (Yenlede *et al.*, 2016).

The aubergine is also believed to have magical and decorative uses in several countries such as India, Benin and the Ivory Coast.

I. 1.4. Justified systematic position of *Solanum melongena*

(**Source**: **1** = Linnaeus, 1753; **2** = Cronquist, 1981; **3** = APG III, 2009)

- eukaryotes.. **Domain** : Eucarya (2)
- multicellular, autotrophic, cellulose wall... **Kingdom**: Plantae (2)
- vascular plants...................................... **S/regnum** : Tracheobionta(2)
- seed plant... **Phylum**: Spermaphytes (2)
- ovum covered by a carpel .... **S/branching** : Angiosperms (2)
- flowering plants whose plants bear fruit..
  ....Division: Magnoliophyta (2)
- seeds with two cotyledons; leaves with branched veins; flowers type 5........................
  .... **Class**: Magnoliopsida (2)
- five fused/free petals; ovary inferior............ **Subclass**: Gamopetales (2)
- small flowers with free valvate petals .... **Order**: Solanales (3)
- annual or perennial; leaves often alternate and compound, bipinnate, monopinnate or palmate, simple; corolla type 2; five stamens, of which two are small and three large; flower bud with two dehiscence slits..
  ....Superfamily: Herbaceae (2)
- herbaceous plants, shrubs or vines with simple alternate leaves without stipules.........

.... **Family**: Solanaceae (3)

\-   includes some of the most important genera and species including eggplant..................

....**Sub-family**: Solanoideae (1)

\-   includes two genera (Solanum and Jaltomata) which account for about half of the species in the family Solanaceae............................. **Tribe**: Solaneae (1)

\-   quasi-cosmopolitan distribution, 9ᵉ row of rows Angiosperms, leaves alternate to the leaf blade, flowers bisexual, anthers in cymose inflorescences...

....Genre: *Solanum* (1)

\-   leaves alternate, simple pointed with white petals ....

.... **Subgenus**: Leptostemonum (1)

\-   .... **Section** : Melongena (1)

Scientific name: *Solanum melongena* L., 1753

Synonym: *Solanum insanum* L. , 1767 ;

Solanum esculentum Dunal (1813) ;

Solanum incanum auct. non L. Benoit Bock, 2019.

I.2 Insect pests of *Solanum melongena*

I.2.1. Seedling pests

At sowing, the seed and young seedling can be attacked by soil-borne fungi, millipedes, wilt, damping-off, destroying the cotyledons of the seed or the roots of the young seedling, preventing the plant from growing or leaving only a weakened seedling (Ontario, 2021)

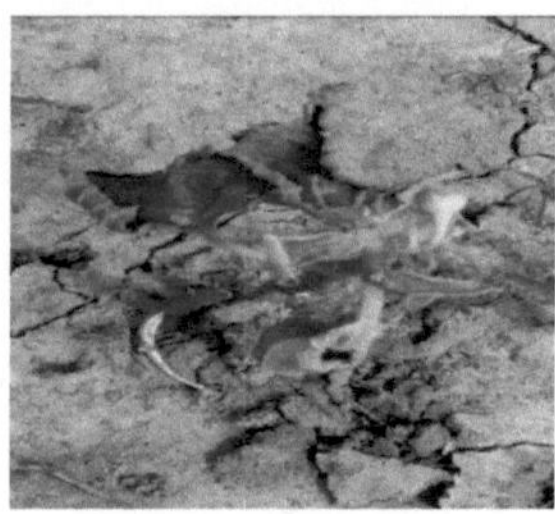

Figure 5: Seedling wilt (*Ralstonia solanacearum*) September 2020

I.2.2. Pests of the vegetative phase

The severity of damage from vegetative phase insects lies in their early onset (Badji, 2010). Generally, these pests contribute to growth retardation, spotting, decay, malformation of leaves, flowers, apices and fruits (Ontario, 2021). Tissue deformation (leaves, stems); flower and fruit drop; loss of germination capacity and reduced growth (Jeanphi, 2017). During this vegetative phase, all parts of the plant are affected (roots, stem, leaves, flowers and fruits). The pests of this phase are divided into 04 pest groups namely:

a. defoliating or phyllophagous species

Figure 6: Some phyllophagous insect pests of *Solanum melongena* (Bocklé, October 2020) (A: Colorado beetle eggs; B: *Acanthocephala femorata*; C: *Haritalodes derogata*; D: *Mantis religiosa*; E: *Arhopalus rusticus*; F: *Tettigonia viridissima*)

b. Stitchers-suckers

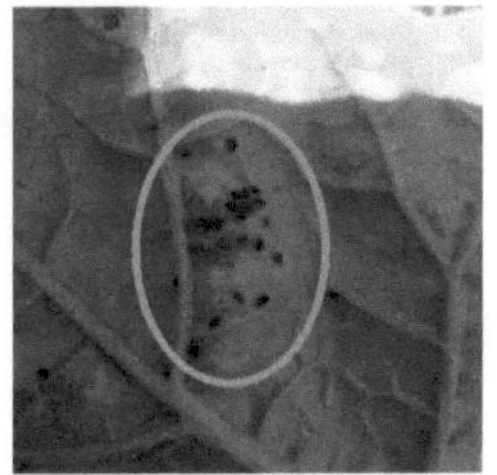

Figure 7: *Aphis gossypii* on a *Solanum melongena* leaf (Bocklé, October 2020)

c. Mites

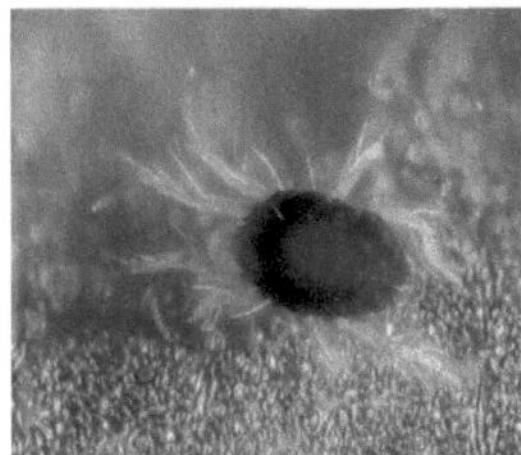

Figure 8: *Tetranychus urticae* on a *Solanum melongena* leaf (Bocklé, December 2020)

d. Root-knot nematodes

Figure 9: Root-knot nematodes (*Meloidogyne* spp.) on *Solanum melongena* roots Guyana, 2019

I.2.3. Pests of the reproductive phase

Figure 10: Some insect pests of the reproductive phase of *Solanum melongena* in Bocklé, November 2020 (A: *Helicoverpa armigera*; B: *Pedridroma saucia* C: *Sarophaga carnaria*; D: *Diogmites neoternatus*)

Insects in the reproduction phase are the main culprits of production losses observed on aubergine. Indeed, in the absence of protection against insect pests, production losses can vary between 25 and 70% (Djibril *et al.,* 2015).

In order to make production profitable, vegetable producers use several methods to control pest attacks. These methods can be grouped into two categories: chemical control and alternative methods to the use of synthetic insecticides.

### I.3	Methods used for the control of pests of *Solanum melongena*

I.3.1. Chemical control

Chemical control consists of the use of chemical products with diverse and specific actions including acaricides, nematicides, insecticides, rodenticides, parasiticides, fungicides, avicides, (Lindquist, 2013). Synthetic pesticides are the most widely used plant protection products by producers due to their availability, ease of use and immediate efficacy (Houadakpode, 2018). They are very effective in controlling pests and are interesting in that their use considerably reduces the drudgery of field work while allowing sufficient production to satisfy both the market and the farmer (Houadakpode, 2018).

However, this control has adverse effects on the environment (Son, 2018), human health (Sola *et al.*, 2014) and also leads to resistance in pests (Agboyi *et al.,* 2018).

I.3.2. Physical struggle

It consists of protecting plants by grouping them by control technique whose primary mode of action does not involve any biological, biochemical or toxicological process, thanks to active and passive methods (Panneton *et al.,* 2012). The former consumes energy at the time of application and the latter modifies the environment and is more sustainable (Panneton *et al.,* 2012). As insect control techniques, we have electromagnetic radiation (microwave, radio frequency, infrared) and pneumatic control (blowing/sucking), thermal and mechanical shocks which are active methods (Houadakpode, 2018). Passive methods applicable in the field are physical barriers including the creation of hedges between plots, pre-englazed yellow traps, anti-insect nets, trapping with specific pheromones and the implementation of appropriate cultural techniques to prevent outbreak problems (Martin *et al.* , 2014).

I.3.3. Crop association or intercropping

It is a practice that consists of growing more than one species on the same plot during a significant period of their growth (Laurent, 2017). It involves two types of ecological mechanisms: a top-down effect and a bottom-up effect.

The top-down effect is the influence of the higher trophic level on the lower trophic level. It is the regulation by predation, parasitism, etc. influenced by the presence of the associated crop (Houadakpode, 2018). Indeed, the associated crop may be a habitat for pest beneficials or for prey. Generally speaking, according to the so-called natural enemies hypothesis (Damien, 2018), natural regulation through parasitism, predation is more effective in complex systems than in simple systems.

Bottom-up effect: this is the influence of the lower trophic level on the higher trophic level (Houadakpode, 2018). In other words, plants in association in a plot have an impact on the behaviour, development and reproductive capacity of insect pests through many visual and olfactory signals (volatile organic compounds or Covs) they emit (Houadakpode, 2018). Indeed, the signals emitted by different plants could also be different. Thus, when plants are in association in the same plot, they emit different signals that could be confusing for insect pests than those emitted in monoculture (Arnaud, 2019).

Aubergine will do well alongside bush beans, string beans, spinach and tarragon, but avoid planting aubergine on a plot that hosted Solanaceae (tomatoes, potatoes) and Cucurbitaceae (melons, cucumbers, squash) the previous season (Paul, 2013). Several studies have shown the effectiveness of crop association in reducing pest populations through repellent or attractive effects (Tiroesele *et al.* , 2015; Yarou, 2018).

I.3.4. Biological **control**

Based on the use by humans of natural enemies such as predators, parasitoids or pathogens to control populations of pest species and keep them below a pest threshold (OILB). For example, against adult whiteflies, thrips, or the leafminer *Tuta absoluta* and *Pedridroma saucia*, the use of beneficials such as *Macrolophus pygmaeus, Amblyeius swirskii* and *Macrolophus caliginosus* are indicated (Laurent, 2014). These control agents only attack specific developmental stages (Nanfack *et al.*, 2015) and their negative effects are becoming rare due to the fact that much effort is being made to learn about their biology and ecology before their introduction (Bouzerida *et al.*, 2016).

I.3.5. Use of plants as insecticides

Plants are an important resource in the life of living beings. It is used in various fields, namely food, medicine, culture and socio-economics. It is also used in agriculture for post-harvest conservation or plant protection (Gakuru *et al.* , 2011). It is a promising alternative for the management of pests of *Solanum melongena* in particular and crops in general while ensuring the preservation of the quality of water, soil, fruits and vegetables for human consumption because bio-insecticides are rapidly degradable in the environment and have a selective effect on living beings (Cloyd, 2004). Their use solves residue problems generated by synthetic insecticides (Ripley *et al.*, 2001; Tano *et al.* , 2011) and they have a less toxic effect on natural enemies of crop pests - in short, they are environmentally friendly and effective in pest control (Stevenson *et al.*, 2014; Mkenda et *al.*, 2015). The use of bio-insecticides is an ancestral practice in Africa and is becoming more widespread in the world, and it is in this sense that the national programmes Champs Ecoles, Biosavane, Écophyto/ ÉcophytoDOM, support actions aimed at moving towards less use of synthetic pesticides in agriculture. A comparative study of the fungicidal effect of essential plant oils and conventional fungicides showed that the essential oil of *Ocicum gratissimum* inhibited spore germination and mycelium growth of *Fusarium oxysporum, Radicis lycopersici, Pythium sp.* and was almost as effective as conventional fungicides (Doumbouya *et al.*, 2012). Various plant species produce chemical substances that are repellent, anti-palatable and even toxic to pests when present in large quantities (Houadakpode, 2018). Insecticidal plants are used in various forms: organic or aqueous extracts of the plants sprayed on the leaves (Nukenine *et al.*, 2013; Mondédji *et al.*, 2014a), in combination with other crops (Bediako *et al.*, 2010; Baidoo et *al.*, 2012) or in the form of essential oils or whole plants for storage of commodities (Nukenine *et al.*, 2010, 2011; Anjarwalla *et al.*, 2016). For our study, we used *Azadirachta indica, Carica papaya* and *Senna didymobotrya*.

I.3.5.1. *Azadirachta indica*

I.3.5.1.1. General information

With neem or free tree as its generic name, it is a majestic leafy plant native to the Indian subcontinent belonging to the Meliaceae family whose fruits and seeds are the source of neem oil.

It is cultivated in tropical and semi-tropical areas (Baptiste, 2016). It is a monoecious tree with usually isolated and compound leaves (Jonathan, 2017). The branches are broad with evergreen foliage that can become deciduous during prolonged drought. Its leaves are opposite and pinnate with dark green leaflets of which the terminal one is absent. Its white, fragrant protandrous flowers appear from January to May in the tropics. Its fruits are an olive-like drupe measuring between 1.4 and 2.8 cm with a thin exocarp skin, a bitter-sweet pulp and a hard white endocarp stone (Lisan, 2012). **I.3.5.1.2. Uses**

Neem is known for its many uses, notably in traditional medicine where its oil extracted from the seeds is used as a powerful disinfectant in Ayurvedic medicine and as a contraceptive (Baptiste, 2016). Its therapeutic qualities, insecticides, insect repellents, vermifuges, anti-venoms, antiseptics and antibiotics help in the treatment of infectious diseases such as leprosy or gonorrhoea, malaria (Lisan, 2012; Baptiste, 2016). Everything is good about it - its fruits, leaves, bark, seeds and roots are also part of the therapeutic arsenal. Similarly, neem is a very powerful insecticide as it is active against more than 400 insects and has been used for centuries to protect crops as well as granaries (Agroecology, 2019). Its seeds are also used to produce a formidable insecticide with azadirachtin as its active ingredient, which disrupts the feeding of the pest, blocks its metamorphosis from the larval to the adult stage (Wondafrash *et al.* , 2012) but also prevents them from eating crops or laying their eggs (Agroecology, 2019).

In addition to its medicinal and insecticidal properties, it has also shown considerable potential as a fertiliser through the solid residue after pressing as it is incorporated into the soil, protecting plant roots from nematodes and white ants. Neem oil is used in the preparation of cosmetic and oral products (GIESS, 2016) and can be lethal if consumed in high doses.

I.3.5.1.3. Systematic position APG 2006

- Pluricellular, autotrophic with a pectocellulosic membrane, they have sexual or asexual reproduction.. **Kingdom**:   Plant
- underground root, vascular plant, non-lignified tissue, comprising a cycle with alternating gametophyte   and   saprophyte...................................... **Division**:   Tracheophyta
- presence of several carpels forming several closed ovaries containing ovules; leaves with branched   veins,   one   or   more   fruits   containing   one   or   more seeds... **Class**:   Magnoliopsida   -

superect ovary with marginal placentation containing two or more anatropic or ompylotropic, gynoecium monocarpellate, the endroecium usually diplostemone............ .... **Order:** Sapindales

- very variable plants, annual or perennial herbaceous, alternate leaves primitively imparipinnate, inflorescence varied (raceme, spike, glomerule), flowers hermaphroditic............. ....Family: Meliaceae

- leaves composed of 3 to 6 almond-shaped leaflets, of a beautiful green colour, the bilaterally symmetrical flowers with 5 rounded golden yellow petals, 2 large curved stamens...................... .... **Genre**: *Azadirachta*

-.. **Species**: *Azadirachta indica*

Zoological Name: *Azadirachta indica*

Synonym: *Largestroemia speciosa, Melia azedarach* A. Juss, 1830.

I.3.5.2. *Carica papaya*

I.3.5.2.1. General

The *pawpaw* (*Carica papaya* L.), commonly known as the tropical melon, is a fruit tree considered to be a large, fast-growing but short-lived (4-7 years) arborescent perennial. It is a dicotyledonous, usually dioecious and sometimes unbranched hermaphrodite plant with evergreen parasol-like foliage in humid and sub-humid tropical areas (Bakahi, 2019). It is a fruit tree native to southern Mexico (Pinnamanneni, 2017) and ranges in height from 5-7m. It is a berry with a leathery pericarp, called a peponid, of various shapes and sizes, appearing directly on the trunk, with yellow or orange or even red pulp and blackish seeds.

I.3.5.2.2. Systematic position APG 2006

- Pluricellular, autotrophic with a pectocellulosic membrane, they have sexual or asexual reproduction... **Kingdom**: Plant

- underground root, vascular plant, non-lignified tissue, comprising a cycle with alternating gametophyte and saprophyte.. **Division**: Tracheophyta

- presence of several carpels forming several closed ovaries containing ovules; leaves with branched veins, one or more fruits containing one or more seeds. .... **Class**: Magnoliopsida

- superect ovary with marginal placentation containing two or more anatropic or ompylotropic, gynoecium monocarpellate, the endroecium usually diplostemone............ .... **Order**: Violales

- very variable plants, annual or perennial herbaceous, alternate leaves primitively imparipinnate, inflorescence varied (cluster, spike, glomerule), flowers hermaphroditic.............
.... **Family**: Caricaceae

- leaves composed of 3 to 6 almond-shaped leaflets, of a beautiful green colour, the bilaterally symmetrical flowers with 5 rounded golden yellow petals, 2 large curved stamens......................
.... **Genre**: *Carica*

-... **Species**: *Carica papaya*

Zoological Name: *Carica papaya*

Synonym: *Papaya carica Gaertn, Papaya edulis Bojer* Linnaeus, 1753.

I.3.5.2.3. Uses

When ripe, papaya is eaten fresh for its nutritional properties with lime or in fruit salads. It has medicinal, cosmetic and even ornamental properties (Bakahi, 2019). Papain is recommended as an enzyme replacement for ulcers. It also cures many other diseases in traditional medicinal uses such as fever, constipation, swollen testicles, tooth decay, dysentery, lower abdomen, dermatitis, intestinal worms, hypertension and diuretic (Allan & MacMillan, 2017). Its fruit is edible, contains a proteolytic enzyme (papain) that is highly valued in cosmetic industries, for the manufacture of softeners, refreshing, regenerating and exfoliating agents (Zubiria, 2021). *Carica papaya has* been used as a bio-insecticide in the work of Ahmad (2018); Wahyuni (2019) against aubergine pests by Mochiah *et al.* (2011); okra and cabbage by Gnago *et al.* (2019).

I.3.5.3. *Senna didymobotrya*

I.3.5.3.1 General

Commonly known as senna, candelabra tree (due to the shape of its inflorescences) or Cassia popcorn due to its popcorn-like buds in bloom, *Senna didymobotrya is a* species of dicotyledonous plant in the family Fabaceae, subfamily Caesalpinioideae. Native to Africa, where it can be found across the continent in many types of habitats. *Senna didymobotrya is a* fast-growing shrub (3 m in 2 years) that can grow up to 9 m in its native range (tropical Africa); its fruit is a flat brown pod up to 12 cm long, which contains up to 16 bean-like seeds, each 1 cm long (Angelis, 2021). This senna is not very hardy (USDA zone 10-12) and does not tolerate long sub-zero temperatures.

I.3.5.3.2. Uses

Senna didymobotra is commonly used as a medicinal plant, notably in the treatment of abdominal pain, diarrhoea, malaria although it has a febrifuge action, jaundice, ringworm, abscesses of the skeletal muscle and its odour has a repellent effect on bees. Also used as a purgative, green manure, ornamental plant *Senna didymobotra* can be used against all kinds of gynaecological diseases and as a poison if used in high doses. It has been used as a bio-insecticide against *Acanthoscelides obtectus* in Bungoma-Kenya (Mining, 2014); against *Caryedon serratus* groundnut pest in Senegal (Cheikh *et al.,* 2015).

I.3.5.3.3. Systematic position APG 2006

- Pluricellular, autotrophic with a pectocellulosic membrane, they have sexual or asexual reproduction... **Kingdom**: Plant
- underground root, vascular plant, non-lignified tissue, comprising a cycle with alternating gametophyte and saprophyte...................................... **Division**: Tracheophyta
- presence of several carpels forming several closed ovaries containing ovules; leaves with branched veins, one or more fruits containing one or more garines... **Class**: Magnoliopsida
- superect ovary with marginal placentation containing two or more anatropic or ompylotropic, gynoecium monocarpellate, the endroecium usually diplostemone............
.... **Order**: Fabales
- very variable plants, annual or perennial herbaceous, alternate leaves primitively imparipinnate, inflorescence varied (cluster, spike, glomerule), flowers hermaphroditic.............
.... **Family**: Fabaceae
- leaves composed of 3 to 6 almond-shaped leaflets, of a beautiful green colour, the bilaterally symmetrical flowers with 5 rounded golden yellow petals, 2 large curved stamens.......................
.... **Genre**: *Senna*
-... **Species**: *Senna didymobotrya* Scientific name: *Senna didymobotra*
Synonym: *Senna occidentalis*; *Cassia occidentalis* H.S & Barnerby, 1982.

A   B   C

Figure 11: Feet of *Carica papaya* (A), *Azadirachta indica* (B) and *Senna didymobotrya* (C) Djamboutou, août 2020

I.4   Pollination

I.4.1   General

Pollination is the mode of sexual reproduction of plants based on the transfer of pollen from the male organ of a flower (stamen) to the female organ (pistil) of the same or another flower (MTES, 2017); it can be either autogamous or allogamous. These two categories of pollination occur in three modes: anemogamy, hydrophily and zoogamy. Anemogamy and hydrophily, which correspond respectively to pollen transport by wind and water (Lautenbach *et al.*, 2012). While zoogamy refers to the transport of pollen by animals in order to fertilise a distant ovule (Francois, 2016). Several authors claim that the presence of insufficient numbers of pollinating insects during the flowering period can negatively affect agricultural production (Tchuenguem, 2005; Project, 2015; Paris, 2016). However, more than 75% of plant species are pollinated by animals, mostly by insects, which play a very large role in pollination. There are nine pollination syndromes in zoogamy distributed in particular in Lepidoptera, Coleoptera, Diptera, Hymenoptera, Birds and some Mammals. The order Hymenoptera and mainly the Superfamily Apoidea plays the most important role in pollination (Vaissière, 2016). Bees have also largely contributed to the evolution of plant species by allowing inter-specific and inter-genetic crosses (Vaissière *et al.*, 2005).

I.4.2 Pollination and plant production

Several studies aim to quantify the economic value of pollinators for agriculture (Vassière, 2016). More than 75% of crops depend strongly or totally on animal pollination. In Cameroon, as well as all over the world, some Apoids have shown their pollinating activity on several plants: *Xylocopa olivacea* on *Phaseolus coccineus* (Tchuenguem *et al.*, 2012; 2014); *Apis mellifera* on *Bracharia brizantha* (Adamou & Tchuenguem, 2014) and on *Physalis minima* (Djakbé *et al.*, 2017) *Dactylurina staudingeri* on *Psorospermum febrifugum* (Zra *et al.*, 2020). Studies have also shown that agricultural intensification threatens bee populations and their pollination services (Alexandra, 2018). Therefore, it is important to promote the maintenance and development of pollinators in order to support not only agriculture, but also plant biodiversity (Ndayikeya & Nzigidaher, 2017).

I.5. *Apis mellifera*

I.5.1 General

The bee is the vernacular name given to hymenopteran insects of the superfamily Apoidea (Hardy, 2012). There are at least 20,000 species of bees worldwide, which are thought to be more than

close to ants than to wasps according to the principle of parsimony (INRA, 2014).

Not all bees are honey producers except those belonging to the genus *Apis* and are divided according to their lifestyle into several groups: honey bees, wild bees, solitary bees and social bees (Bellman, 2009). A social bee society has three castes according to their functions and structures: the queen, the workers and the false bees (Aymé, 2014).

The queen has a long body with a well-developed abdomen, can lay up to 2,000 eggs per day and can live up to 5 years (Aymé, 2014). Males are stocky, from unfertilised eggs, with a large abdomen that does not extend beyond the wings (Aymé, 2014). They are not always present in the hive and are responsible for mating with the queen during the mating flight (INRA, 2014). Workers are the smallest bees in the hive with wings the same length as the abdomen and have anatomical adaptations related to their activity, also a waxing apparatus and pollen basket on the hind legs (Aymé, 2014).

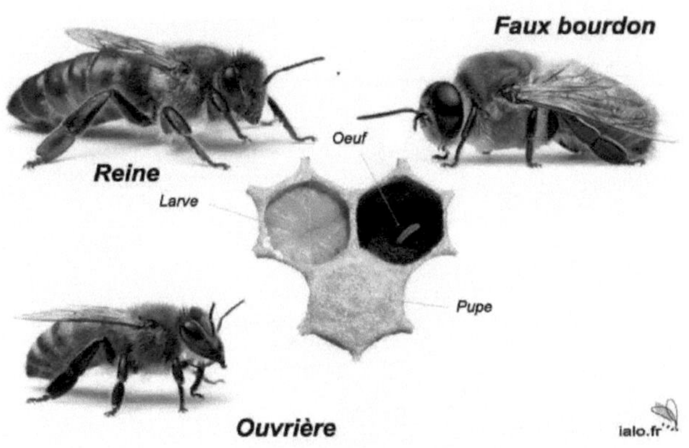

Figure 12: Different castes in an *Apis mellifera* colony (Helden, 2019)

The bee that has found and exploited a source of nectar arrives at the hive, regurgitates part of its nectar harvest, and then immediately, on the comb where it is, performs a series of very stereotyped movements called "dances" (Frisch, 1984; Maurice, 1968). In the "round dance", it describes a circular trajectory, returns to its starting point, turns around and repeats the same movement in the opposite direction (Frisch, 1967). The round dance (Figure 13) only concerns spoils that are located within about 100 metres of the hive (Frisch, 1967).

When the food source is further away from the hive, the scouting bee that has come to warn her companions performs a more complicated dance to indicate the direction and distance of the place to be found: this is the wriggling dance (Figure 13), which can be observed in its most typical form as soon as the distance is at least 100 metres (Frisch, 1967). On the comb, the bee first makes a short straight path (crossing a maximum of five rows of cells, in the case of long distances); it then describes a semicircle which brings it back to its starting point, repeats the straight path, describes a new semicircle symmetrical to the first one (Frisch, 1967). The Girl Scout then repeats this complete journey for a few minutes (Frisch, 1967). With each straight path, the dancer begins to wriggle her abdomen, i.e. to vibrate it rapidly from left to right while emitting a rhythmic buzzing sound (Frisch, 1967). The alert workers then follow the dancer and feel her with their antennae to receive various messages (Frisch, 1967; Maurice, 1968).

In nature, the bee collects five essential products for feeding and habitat protection: water, nectar, honeydew, pollen and resin (Morison *et al.,* 2000).

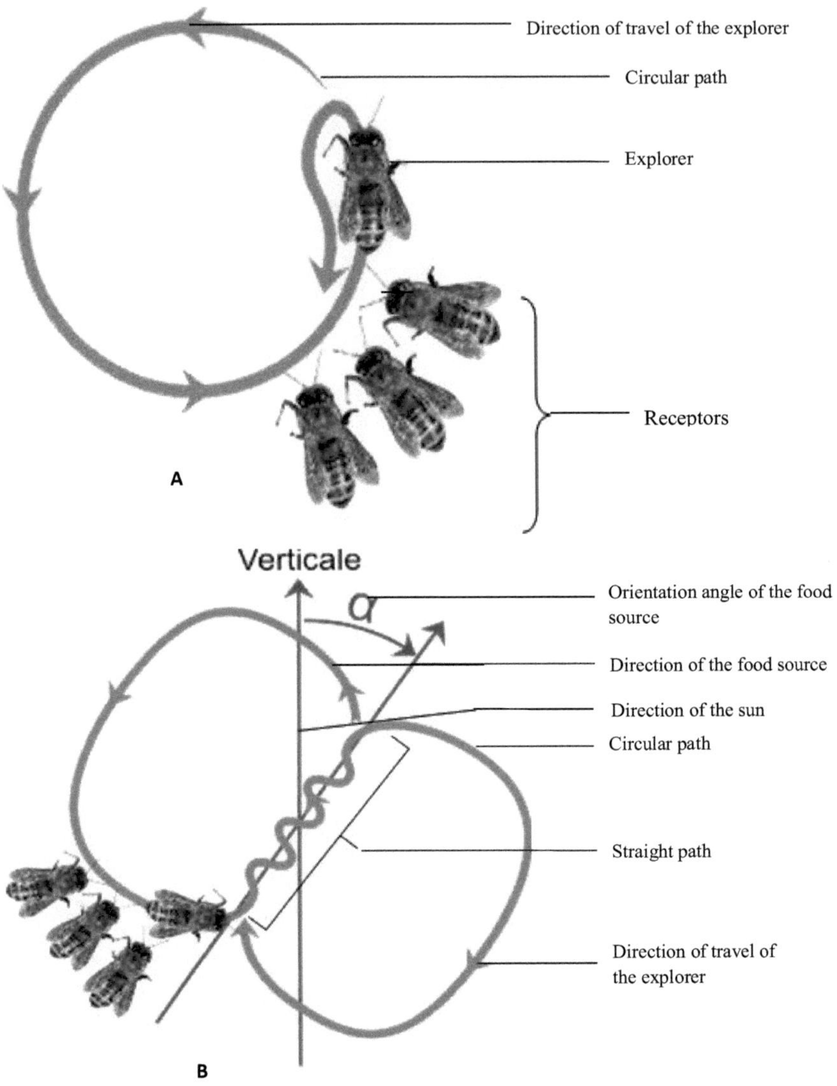

Figure 13: Diagram of the round dance (A) and the wriggling dance (B) of an *Apis mellifera* explorer (Frich, 1967)

I.5.2. Systematic position of *Apis mellifera*

(**Sources: 1** = Woese et al., 1990; 2 = Borror & White, 1991; 3 = Roth, 1980; 4 = Grassé, 1951; 5 = Michael, 1999).

-eucaryote...................... ....Domain: Eucarya (1)

-multicellular ;   heterotrophic;   no   cell   wall   ....
.... **Kingdom**: Animalia (1)

-metamerised body; chitinous cuticle; articulated appendages.......................................
....Branch : Arthropods (2)

-antennae   ; mandibles.......................... **Subphylum**: Mandibulates (2)

-tracheal respiration ; body divided into head, thorax and abdomen; three pairs of legs..Class: Insects (2)

-wings... **Subclass**: Pterygotes (2)

-Original wing observation....................... **Super - order**: Hymenopteroids (3)

-two   pairs of sparsely-veined membranous wings, the forewings coupled to the hindwings by a clasping device; forewings larger than the hindwings; tarsi with five articles; mouthparts crushing; maxillae and labium modified into a sucking tongue; antennae with at least   10 articles; holometabola......... ....
....Order: Hymenoptera (2)

-Petiole abdomen   ; thorax with 4th segment; propodeum (1st abdominal segment) fused to thorax...................................... **Sub-order**: Apocrites (2)

-pronotum   short, collared; 1st article of hind tarsi enlarged and flattened...................
....Super - **family**: Apoidea (2)

-forewings with three submarginal cells; two basal articles of labial palps long and flat; maxillary palps reduced; large, long and thin ....
.... **Family**: Apidae (2)

-queen   larger than the worker, with reduced harvesting and building organs and a medium-sized sting; pure wax cakes, hanging from branches or rocks; hives sometimes densely populated; swarming of the old founding queen or of a young female surrounded by workers.................
.... **Sub-family**: Apinae (4)

-builders   of vertical combs made of wax; presence of a communication system in the form of "dances"; social insects with three morphologically distinct castes; new colonies founded by fission and including the former queen.............. ....
....Genus: *Apis* Linné **(5)**

-        nest in a cavity; "dances" performed on a vertical surface of a comb; during the "dances", abdomen wriggling parallel to the surface of the comb and recruits positioned close to the abdomen of the "dancer"......................... **Subgenus**: *Apis* Linnaeus **(5)**

-        hairy eyes; hind tibiae without apical spurs; narrow, parallel-sided forewing marginal cell; oblique 3rd marginal cell..

.... **Species**: *Apis mellifera* Linné **(2)**

Scientific name: *Apis mellifera* Linnaeus, 1758 **(5)**

Synonym: *Apis mellifica* Linnaeus, 1761 **(4)**

I.6   Relationship between *Solanum melongena* and floricultural insects

The floristic entomofauna and the impact of insects on pollination and fruit and/or grain yields of a plant species can vary with insect species, time and space (Tchuenguem, 2005). Aubergine flowers are perfect and self-fertilising. However, inter-varietal hybridisation can occur and the frequency varies according to the environment and the nature of the insect pollinators. Some studies in southern India have shown that aubergine flowers are regularly visited by butterflies, mason wasps and solitary bees (Kokopele, 2013); Fayet in France has shown that bumblebees and xylocopes are the most suitable insects for pollinating aubergine flowers.

I.7.   Relationship between *Solanum melongena* and pests

The relationship between aubergine and insect pests is little studied. Parasites (mites, aphids, leafminers, nematodes, leafminers, bugs and moths) are an obstacle and a real source of constraint for aubergine because they cause multiple diseases from sowing to harvesting that considerably affect the yields of this plant (PlantPathology, 2014).

I.8.   Relationship between *Solanum melongena*, aqueous plant extracts and floricultural insects

To our knowledge, there are no publications dealing with the cumulative impact of botanical insecticides and floricultural insects on the yields of *Solanum melongena*.

CHAPTER II: MATERIALS AND METHODS

II.1 Presentation of the experimental site

Our study was carried out in a garden in Garoua (Bocklé) which is the capital of the North Cameroon region and the chief town of the Benue department. The city of Garoua is located between 9e degrees, 18 minutes, 5 seconds North Latitude and 13e degrees, 23 minutes 52 seconds East Longitude with an Altitude of 249m (Figure 14).

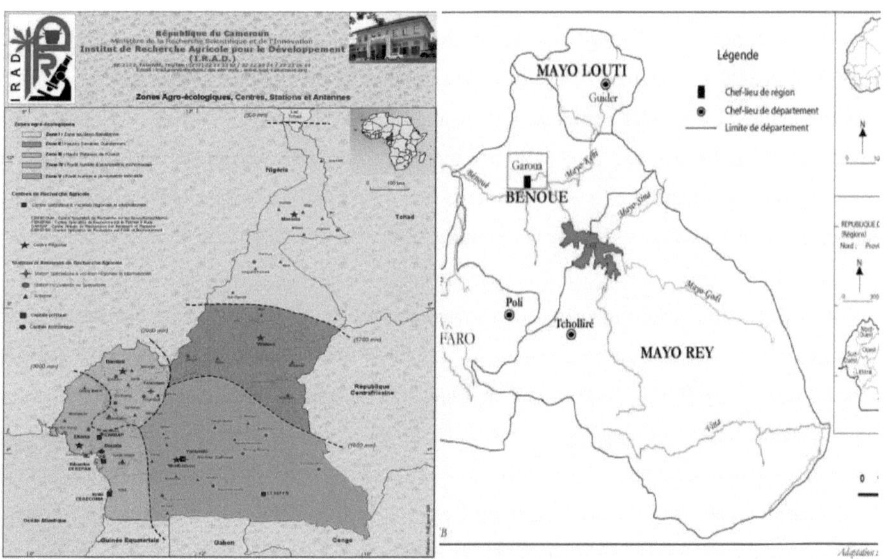

Figure 14: Map of Cameroon showing the locality of Garoua, the different agricultural research centres and branches, the main agro-ecological zones (IRAD, 2016).

II.1.1. Human, geographical, climatological and vegetation overview of Garoua

The population of Garoua, estimated at 5,460,460 inhabitants, is distributed among the ethnic groups according to the following order of sedentarisation: the indigenous people (Falis and Batas), the Peuls, the Haoussas, the Kanouris, the Lakas, the Mboums, the Massas, the Toupouris, the Moudangs and the ethnic groups resulting from recent emigration (Bidzoni, 2020).

The city is an oasis of greenery in the heart of the bush, Garoua is the only city in Cameroon crossed by the Benue River. It is a pleasant and well-equipped city that is a hub for tourist movements directed, on the one hand, towards the Far North via the Waza National Park and the Kapsiki region, and on the other hand, towards the South via the Benue and Boubandjida parks and the Faro reserve on the road to Ngaoundéré (Bidzoni, 2020).

Garoua has a tropical savannah-Sudanese climate with a rainy season that is hot, oppressive and overcast and a dry season that is hot and partly cloudy. Over the course of the year, the temperature generally varies from 18°C to 40°C and is rarely below 16°C or above 45°C. There is considerable seasonal variation in the percentage of cloud cover throughout the year, with a clearer period starting around the end of October with 46% overcast and a cloudier period starting around mid-March with 81% overcast. Garoua experiences extreme seasonal variations in monthly rainfall. The rainy period of the year lasts 7 months with an average total accumulation of 197 millimetres and the dry period of the year lasts 5 months with an average total accumulation of 0 millimetres. The city experiences extreme seasonal variations in perceived humidity including a heavier period with 25% and a lighter period with a heavy climate at 1%. Instantaneous wind speed and direction vary more than the hourly averages with a very windy period (average wind speed over 9.5 kilometres per hour) and a calmer period (hourly average wind speed of 6.1 kilometres per hour) coming from the west, east and north (Weatherspark, 2019).

The vegetation is formed of wooded savannahs and trees whose domain extends from the Adamaoua plateau in the south to the Mandara Mountains in the north, interspersed with Sahelian elements formed by thorny steppes and grasslands. The species encountered here are *Acacia caffra* (acacia), *Afzelia africana, Cassia sieberiana, Combretum, Adansonia digitata* (baobab), *Borassus aethiopum* (bramble), *Bombax costatum* (Kapokier), ficus and tamarind.

Garoua's economy is based on import-export, agro-pastoral activities, forestry, hunting, handicrafts, industry (SABC, ENEO, SAHEL, SPRINT, LANAVET, IRAD), services (hotels), tourism, transport, trade, banks, energy, cultural infrastructures (mosques, churches, telecommunications, media) (Bidzoni, 2020).

II. 1.2 Study station

The investigations were carried out from 1er August to 22 December 2020 in Bocklé in an experimental field 26 m long and 19 m wide set up for this purpose, i.e. an area of 494 m^2 . This plot is centred on a point with the following geographical coordinates: latitude: 9.29255; 9°17'33.17665"N; longitude: 13.41838; 13°25'6.15245"E; altitude: 167 m. These coordinates were taken using a GPS application installed in a "HUAWEI mobile phone, Model mate 9".

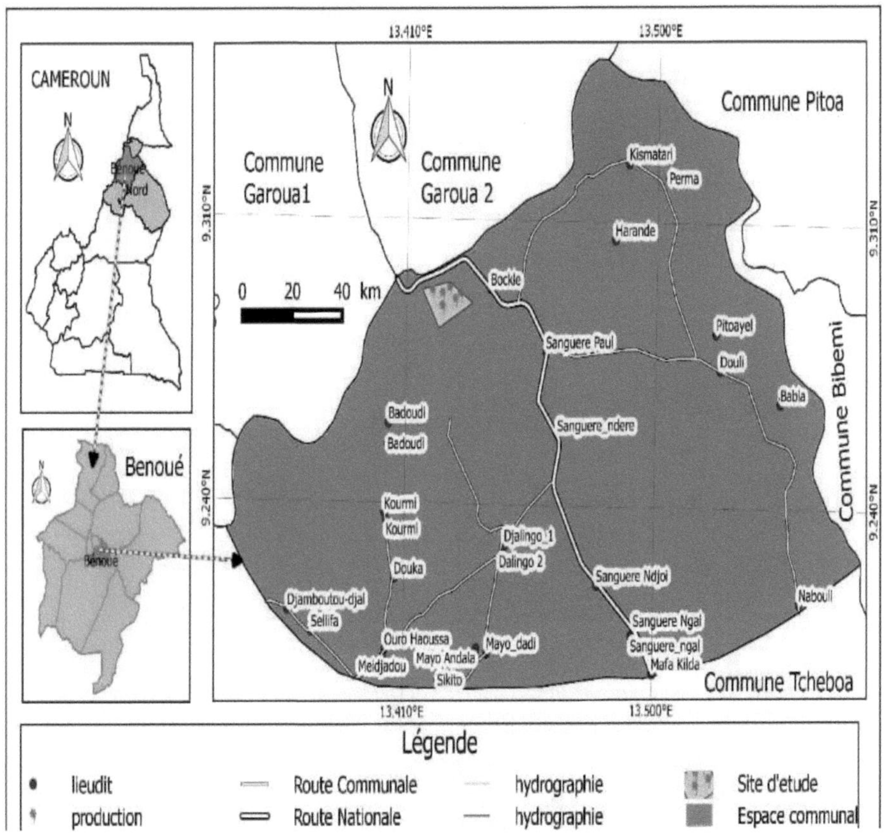

Figure 15: Location map of the study site SOGEFI, 2018

II.2 Materials

II.2.1. Animal material

The animal material consisted of all the insects present on the experimental site and the bee *Apis mellifera* which had several possible origins:

- four colonies in the experimental site; these colonies were housed in various types of habitats: on a street lamp, in a roof of a house, in sovereign cavities, on tree branches (Figure 16);

- other uninventoried colonies in the vicinity of the experimental site, since the range of the foragers can exceed 12 km around the hive (Louveaux, 1984).

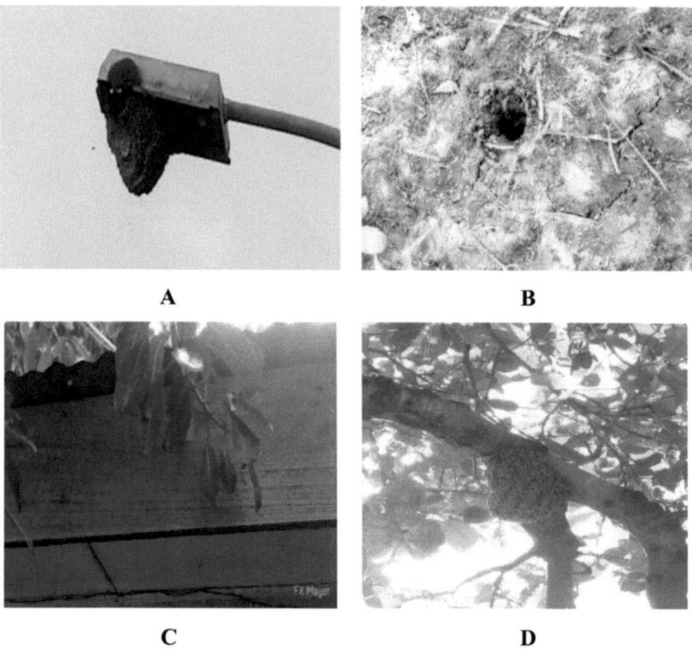

Figure 16: Some *Apis mellifera* habitats of the study station each hosting a colony during the flowering of *Solanum melongena* in Bocké, 2020

A: Lampadaire sheltering a colony of *Apis mellifera*; **B**: Underground cavity sheltering a colony of *Apis mellifera*; **C**: Roof of a house sheltering a colony of *Apis mellifera*; **D**: Branch of *Anacardium* sheltering a colony of *Apis mellifera*.

II.2.2. Plant material

It consisted of seeds of *Solanum melongena* white variety bought in a shop selling agricultural seeds in Garoua (SEMAGRI) and dry leaves of *Azadirachta indica* (Meliaceae), *Carica papaya* (Caricaceae) and *Senna didymobotrya* (Fabaceae) (Figure 17) Leaves were collected frequently at 02 week intervals in Djamboutou.

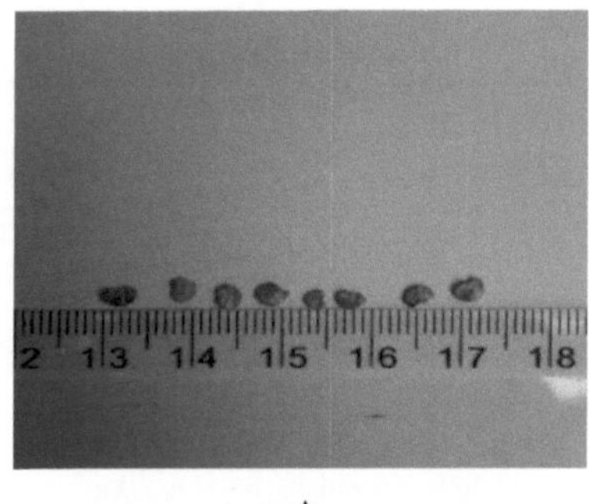

A

B   C   D

Figure 17: Seeds of *Solanum melongena* (**A**), leaves of *Azadirachta indica* (**B**)*, Carica papaya* (**C**) and *Senna didymobotrya* (**D**)

II.3 Methods

II.3.1. Preparation of the experimental field

From 1er to 7 August 2020, a 494 m^2 rectangular plot of land was delimited, cleared, ploughed, fenced and divided into 4 blocks containing 5 sub-plots each, i.e. a total of 20 experimental units of 4 m in length and 3.5 m in width, each separated from the others by a 1 m path.

The regular distance between the patches was measured using a decameter with strings attached and the field was laid out in a complete randomised block design (or RCBD) with nine treatments: three plant extracts, a synthetic insecticide control, the untreated control where each treatment was replicated four times, and four pollinator treatments (Figure 18).

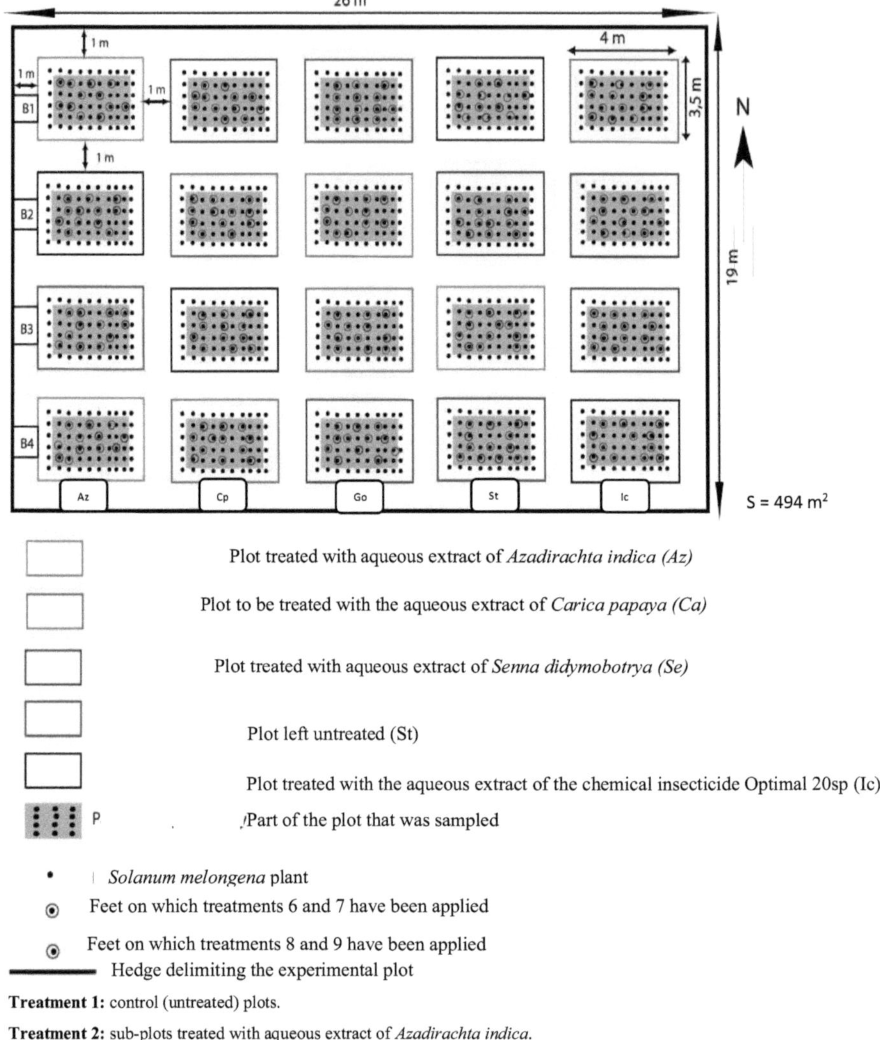

Plot treated with aqueous extract of *Azadirachta indica (Az)*

Plot to be treated with the aqueous extract of *Carica papaya (Ca)*

Plot treated with aqueous extract of *Senna didymobotrya (Se)*

Plot left untreated (St)

Plot treated with the aqueous extract of the chemical insecticide Optimal 20sp (Ic)

P   /Part of the plot that was sampled

- *Solanum melongena* plant
- ⊙ Feet on which treatments 6 and 7 have been applied
- ⊙ Feet on which treatments 8 and 9 have been applied
- —— Hedge delimiting the experimental plot

Treatment 1: control (untreated) plots.

Treatment 2: sub-plots treated with aqueous extract of *Azadirachta indica*.

Treatment 3: sub-plots treated with aqueous extract of *Carica papaya*.

Treatment 4: sub-plots treated with aqueous extract of *Senna didymobotrya*.

Treatment 5: under plots treated with the synthetic insecticide Optimal 20sp.

Treatment 6: 120 flowers at bud stage labelled and left to pollinate freely.

Treatment 7: 120 flowers at bud stage labelled and protected from insects with gauze bags.

Treatment 8: 200 flowers at the bud stage labelled and protected, then destined for the visit of *Apis mellifera*.

Treatment 9: 100 flowers at the bud stage labelled and protected, then intended for opening and closing without visit from insects or any other organism.

Figure 18: *Solanum melongena* experimental set-up (Bocklé, 2020)

II.3.2. Nursery sowing, transplanting and crop maintenance

The seeds of *Solanum melongena* were broadcast on 04 August 2020. The soil was then stirred, watered and covered with straw to maintain moisture to facilitate seed emergence. The nursery thus set up took 18 days before being dismantled and then transplanted (22 August) with two vigorous plants per pot. The transplanting was done in rows with ten rows per sub-plot. Spacing within and between rows was 50 cm and 36 cm respectively (Marthe, 2016; Kingha, 2014) and is illustrated by the partial view of the *Solanum melongena* field 45 days after transplanting at Bocklé in 2020 (Figure 19).

Figure 19: Partial view of the *Solanum melongena* experimental field 45 days after transplanting at Bocklé in October 2020

II.3.3. Pest control

II.3.3.1. Formulation of insecticide products

II.3.3.1.1. Aqueous extract

Fresh leaves of *Azadirachta indica*, *Carica papaya* and *Senna didymobotrya were* harvested regularly at two-week intervals from September to November 2020 in Djamboutou, North Cameroon Region. These leaves were dried and then ground into powder using a mortar and stored dry in plastic boxes.

The aqueous extract based on the plant powders was obtained by the modified method of Anjarwalla & Stevenson (2016). 125 g of each of these powders were weighed using an electronic balance (Ponenlie SF400) and then mixed with 1.5 L of water and left to stand for at least 24 h for better maceration. After maceration, the mixture was filtered and diluted to 10% with water and put in a hand sprayer before use in the field (figure 20).

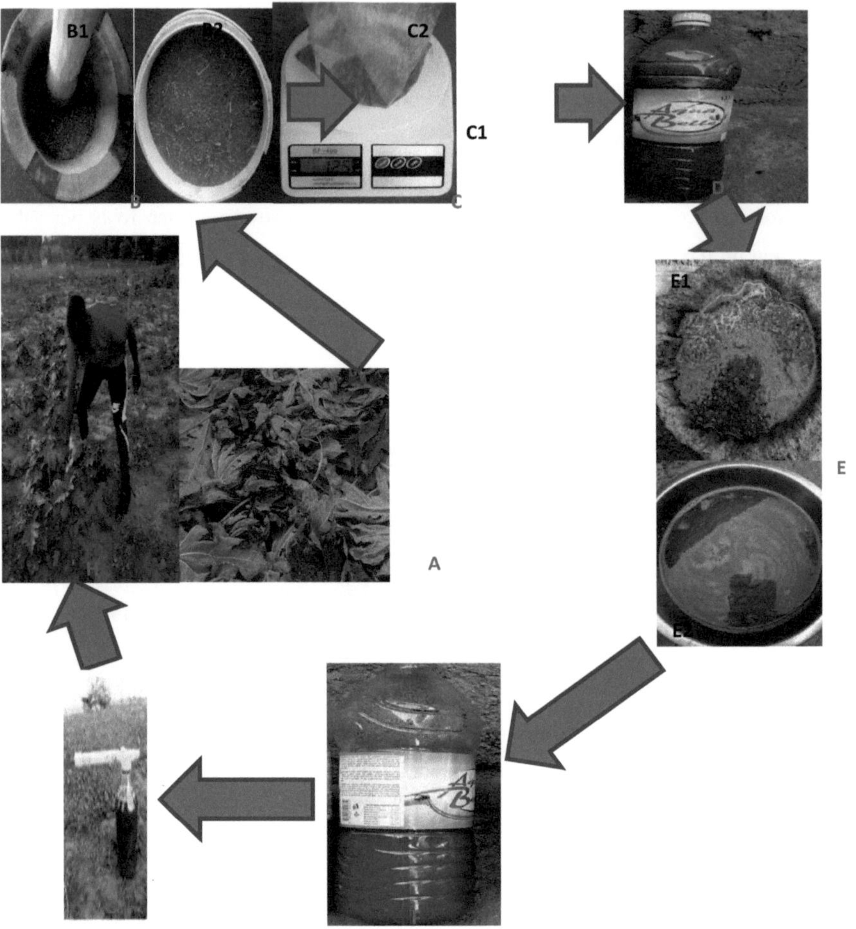

Figure 20: Process for extracting botanical insecticides from dry plant leaves plants

A: Drying; **B**: Grinding (**B1**: Mortar + pestle; **B2**: Can); **C**: Weighing (**C1**: Ponelie SF400 electronic scale; **C2**: Insecticide powder); **D:** Maceration; **E**: Sieving (**E1**: Filtration; **E2**: Filtrate); **F**: Filtrate before dilution; **G**: Solution before use; H: Field spraying

II.3.3.1.2 Chemical insecticide

The chemical insecticide formulation followed the labelled dosage of 5g of Optimal 20 SP synthetic insecticide powder diluted in 1.5L of tap water (Figure 21).

Figure 21: Optimal 20 SP synthetic insecticide formulation

A: Bag of Optimal 20 SP powder; **B**: Weighing of 5g of Optimal 20 SP powder to be diluted; **C**: Dilution of 5g of Optimal 20 SP powder in 1.5L of water; **D**: Manual field spraying.

II.3.3.1.3. Application of insecticide products

The spraying of the substances was done using 4 hand-held 1.5L volumetric sprayers. Each sprayer corresponded to one treatment, i.e. *A. indica, C. papaya, S. didymobotrya* and Optimal 20 SP. The aqueous extracts of the plant powders were applied at a concentration of 10% (by adding 0.15L to the filtered macerate). These extracts including Optimal 20 SP were sprayed in the evening at sunset (from 5pm). Spraying was started two weeks after transplanting, and repeated every fortnight until harvest.

II.4. Data collection

Data collection started 2 weeks after transplanting. It was repeated every 7 days and was done every morning (from 6:00 to 10:30). The parameters measured were the number of insect species, leaves, flowers, fruits per plant/subplot and the yield in
seeds.

The population of these insect pests was assessed on the plants in the middle rows of each sub-plot. To do so, the counting was done with a magnifying glass for the smallest insects and with the naked eye for the most visible ones.

The number of leaves, flowers and fruits per plant was assessed by counting these parameters according to the phenology of the plant.

For the yield, the fruits were harvested from each sub-plot with a preference for the middle rows. The dried fruits were opened manually and the weight (in Kg) of the seeds was assessed using an electronic Ponenlie SF400 scale.

For all these parameters measured, 32/60 plants/subplot were sampled. Throughout the investigation period, insect pests were targeted according to the phenology of the plant studied.

II.5. Impact of pollinators

II.5.1. Determination of the mode of reproduction of *Solanum melongena*

As soon as the first flower buds appeared, 240 buds were tagged on 120 *Solanum melongena* plants at a rate of 6 plants per sub-plot and two treatments were set up:

-Treatment 6: 120 flowers left to pollinate freely and on which no insects were caught (Figure 22);

- Treatment 7: 120 flowers protected from insects with 1 mm mesh bags[2] (Figure 23).

Figure 22: Foot of *Solanum melongena* carrying a labelled flower and left to pollinate freely (Bocklé, October 2020)

Figure 23: *Solanum melongena* plant with a labelled flower protected from insects by a gauze bag (Bocklé, October 2020)

At harvest, the number of fruits formed was counted on the plants of each of the treatments 6 and 7. For each treatment, the fruiting index *(Ifr) was* calculated according to the formula below:

Ifr = (F7 / F6) where *F7* is the number of fruits formed and *F6* the number of flowers initially borne (Tchuenguem *et al.,* 2001).

The difference between the fruiting indices of the two treatments was used to calculate the rates of allogamy *(TC)* and autogamy in the broad sense *(TA)*, according to the following formulae (Demarly, 1977):

*TC = {[(IfrX - IfrY) / IfrX] * 100},* where *IfrX* and *IfrY are* the fruiting indices in the free and protected treatment respectively;   *TA = (100 - TC).*

II.5.2. Determination of the place of *Apis mellifera* in the floricultural entomofauna of *Solanum melongena*

Observations were made daily during the flowering period from 27 to 05 October 2020, during six time slots (6 - 7 am, 8 - 9 am, 10 - 11 am, 12 - 1 pm, 2 - 3 pm and 4 - 5 pm), on the flowers of treatment 6.

We passed once over each flower in treatment 6 and during each of the above time slots. At each pass, the different insects encountered on the flowers were identified and counted. As the insects were not marked, the cumulative results were expressed as the number of visits (Tchuenguem, 2005). The data obtained made it possible to determine the frequency of each insect species (Fi) in the floricultural entomofauna of *S. melongena* (Tchuenguem, 2005), using the following formula

$Fi = (Vi / V_I) * 100$, with Vi = number of visits of insect i to the flowers of treatment 6 and V_I the number of visits of all insects to the same flowers (Tchuenguem *et al.,* 2001).

II.5.3. Study of the activity of *Apis mellifera* on the flowers of *Solanum melongena*
II.5.3.1. Visit rate according to time slots and observation days under the effect of insecticide treatments

Data on the relative frequency of *A. mellifera* visits to the flowers in treatment 6 were used.

II.5.3.2. Harvested floral products

The aim was to note how the bee collects pollen from a flower. The bee that scratches the anthers with its mandibles and legs is a pollen gatherer (Tchuenguem, 2005). The pollen collected can be observed on the baskets of the metathoracic legs (Cadenas, 2017). The floral product was systematically noted when recording the duration of visits per flower (Tchuenguem, 2005).

II.5.3.3. Abundance of foragers

It is expressed by the highest number of *A. mellifera* foragers simultaneously active on a flower, and on 1000 flowers (Tchuenguem, 2005) according to the different insecticide treatments. This parameter was recorded during six daily time slots (7 - 8 h, 9 - 10 h, 11 - 12 h, 13 - 14 h, 15 - 16 h and 17 - 18 h), with at least five values per time slot, when bee activity allowed.

II.5.3.4. Evaluation of the relationships between the flowering rhythm of *Solanum melongena* and the visiting rhythm of *Apis mellifera* under the effect of insecticide treatments

The number of labelled open flowers not protected from insects (treatment 6) was counted every day from the beginning of flowering until the last flower faded. The data obtained were compared with the number of visits of *A. mellifera* to the corresponding flowers (Tchuenguem, 2005).

II.5.3.5. Duration of visits per flower

This is the time taken by an *A. mellifera* worker to collect pollen from a flower. This parameter was recorded on the same days and daily time slots as the abundance of foragers according to the different insecticide treatments, with at least five values per time slot, when bee activity permitted.

II.5.3.6. Foraging ethology

The foraging behaviour of *A. mellifera* under the influence of insecticide treatments was noted throughout the study period, by rigorous observation of its activity at the flower level. The foraging rate, which corresponds to the number of flowers visited by an insect in one minute (Decourtye, 2016), was recorded taking into account the effect of the different insecticide treatments. During this observation, when a worker returns to a previously visited flower, the count is made as if it were two different flowers (Tchuenguem, 2005). The data were recorded on the same dates and daily time slots as for the duration of the visits, with at least five readings per time slot.

II.5.3.7. Foraging ecology

a. Influence of wildlife

The influence of fauna on the foraging behaviour of *A. mellifera was* systematically recorded when timing the duration of visits per flower. Interrupted visits were marked with a distinctive sign indicating the cause of the interruption.

b. Influence of the surrounding flora

It was assessed by direct observations. The number of times that *A. mellifera* moved from a flower of *S. melongena* to the flowers of another plant species and vice versa was recorded (Tchuenguem, 2005).

c. Influence of some climatic factors

The temperature and hygrometry of the study station were recorded every 30 minutes from 6:00 a.m. to 6:00 p.m. using a portable branded thermo-hygrometer (Techno Line WS 9119) installed in the shade (Tchuenguem, 2005), throughout the observation period. The effects of wind, sunshine, rain and cloud cover were also noted. The duration of each visit interrupted by wind was marked with a distinctive sign (Tchuenguem, 2005).

II. 6. Estimation of the beekeeping value of *Solanum melongena*

As in other works (Fameni *et al.* , 2012; Dounia & Tchuenguem, 2014; Mazi, 2015; Djakbé *et al.,* 2017; Farda, 2018), the beekeeping value of *Solanum melongena* was assessed using data on :

-the   intensity of flowering ;

-the   attractiveness of *A. mellifera* workers to pollen.

II.7. Evaluation of the impact of *Apis mellifera* on the pollination of *Solanum melongena*

This parameter was recorded when studying the duration of visits per flower. This involved recording, during pollen collection, the number of times the body of *A. mellifera* comes into contact with the stigma or anther of the visited flower (Tchuenguem, 2005; Az'o *et al.,* 2010; Djonwangwé *et al.,* 2011a, 2011b, 2011c; Kingha *et al.,* 2014; Kengni *et al.,* 2015).

II.8. Evaluation of the cumulative impact of insecticide treatments and floricultural insects including *Apis mellifera* on the yields of *Solanum melongena*

It is based on :

- the cumulative impact of anthophilous insects and botanical extracts on pollination;
- the cumulative impact of anthophilic pollination and botanical extracts on fruiting;
- comparison of fruit yields (fruiting rate) and seed yields (average number of seeds per fruit and percentage of normal (well-developed) seeds) from the treatment with flowers left to pollinate freely to those from the treatment with protected flowers, then destined to be opened and closed by insects (Tchuenguem *et al.*, 2001) under the influence of insecticide treatments

The fruiting rate (Pi) due to the influence of flowering insects was calculated using the formula :

$Pi = \{[(F6 - F9) / F6 + F7 - F9] * 100\}$ where F6, F7 and F9 are the fruiting rates in treatments 6 (free flowers), 7 (protected flowers) and 9 (protected flowers then destined for opening and closing without visits from insects or any other organism) respectively. For a treatment x, the fruiting rate (Fx) is :

$Fx = [(\text{number of fruits} / \text{number of flowers}) * 100]$ (Diguir *et al.*, 2020).

The percentage *(P)* of the number of seeds per fruit attributable to the influence of floricultural insects was calculated using the formula :

$P = \{[(g6 - g9) / g\,6 + g\,7 - g\,9] * 100\}$ where 6 (free flowers), 7 (protected flowers) and 9 (protected flowers then destined to open and close without visits from insects or any other organism) respectively (Diguir *et al.*, 2020).

The percentage *(Pn)* of normal seeds attributable to the influence of floricultural insects was calculated using the formula :

$Pn = \{([Pn6 - Pn9) / Pn\,6 + Pn\,7 - Pn\,9] * 100\}$ where 6 (free flowers), 7 (protected flowers) and 9 (protected flowers then destined to open and close without visits from insects or any other organism) respectively (Diguir *et al.*, 2020).

II.9. Estimation of the pollination efficiency of *Apis mellifera* on *solanum meongena* according to the treatments with insecticides

In parallel with the implementation of treatments 6 and 7, 300 flowers were labelled and two treatments were set up:

- **Treatment 8:** 200 flowers labelled at the bud stage and protected from insects like those in treatment 7, then intended for the exclusive visit of *A. mellifera* ;
- **Treatment 9:** 100 flowers labelled at the bud stage and protected and then intended for opening and closing without visits from insects or other organisms.

As soon as each flower in treatment 8 had opened, the gauze was carefully removed during the daily period of optimal insect activity and the flower was left to pollinate freely for one to ten minutes, to note any visits by *A. mellifera*. After this operation, the flower was protected again and not handled. For treatment 9, as soon as each flower had bloomed, the gauze was carefully removed during the daily period of optimal activity of *A. mellifera* and the flower was left to pollinate freely for one to ten minutes, avoiding being visited by an insect or any other organism. After this manipulation, the flower was again protected and not manipulated.

At fruit maturity, harvesting was done in treatments 8 and 9 and the percentage fruiting rate for each treatment was calculated. The fruiting rate, the average number of seeds per fruit and the percentage of normal seeds per fruit attributable to the influence of *A. mellifera* and the manipulation were determined.

The fruiting rate attributable to *A. mellifera* activity *(Pt)* was calculated using the following formula:

$$Pt = \{[(F8+Eg) - F9) / (F8+Eg)] * 100\}$$

where F8 and F9 are the fruiting rates in treatments 8 and 9 respectively and Eg the effect of the gauze bag.

$$Eg = F8 - F9$$

where F9 is the fruiting rate in the treatment with protected reproductive shoots, uncovered and then protected again without visits from insects or any other organism (Tchuenguem, 2020).

The percentage of the average number of seeds per fruit attributable to *A. mellifera* activity *(Pg)* was calculated using the following formula:

$$Pg = \{[(g8 - g9) / g8] * 100\}$$

where *g8* and *g9 are* the average number of seeds per fruit in treatments 8 and 9 respectively (Tchuenguem, 2005).

The percentage of normal seeds *(Pgna)* attributable to the influence of *A. mellifera was* calculated using the following formula:

$$Pgna = [(Pgn8 - Pgn9) / Pgn8] * 100$$

where *Pgn8* and *Pgn9 are* the percentages of normal seeds in treatments 8 and 9 respectively (Tchuenguem, 2005).

II.10. Capture and determination of insects

Throughout the observation period, insect samples, except for *A. mellifera* active on aubergine, were captured (2 to 3 individuals per species) with entomological nets and by hand. In the field, these specimens were preserved in vials containing 70% ethanol, according to the recommendations of Borror & White (1991), with the exception of Lepidoptera which were preserved in papillotes. The determination of these samples was done by ourselves with the help of specialised books, research applications and with the help of Dr ADAMOU Moïse and Prof NUKENINE Elias.

II.11. Collection and determination of plants

During the observation period, samples of the various plants with reproductive shoots that attracted the forager workers were collected and a herbarium was made. The identification of the specimens was done with the help of the teachers of the Faculty of Sciences of the University.

II.12. Data processing

The data were processed using Excel 2016.

The analysis was done using SPSS (Statistical Package for the Social Sciences) software through :

- descriptive statistics (calculation of means, frequencies, standard deviations and percentages);
- four tests: (a) Pearson's correlation coefficient (r) for the study of linear relationships between two variables; (b) Analysis of variances (ANOVA) for the comparison of means of more than two samples; (c) Student's *t-test* for the comparison of means of two samples; (d) chi-square (χ^2) for the comparison of percentages.

CHAPTER III: RESULTS AND DISCUSSION

III.1 Mode of reproduction of *Solanum melongena*

The fruiting index was 0.99 (n = 120; s = 13.27) in treatment 6 and 0.72 (n = 120; s = 15.46) in treatment 7. Thus, TC = 27.27% and TA = 72.73%. Therefore, *Solanum melongena* has a mixed allogamous-self-pollinating mode of reproduction, with preferential self-fertility. These results are in agreement with those published by Kokopelli, (2020).

III.2 Entomofauna of *Solanum melongena*

III.2.1. Pests inventoried on *Solanum melongena*

During the observation period, 12 species of insect pests were counted with 7208 individuals on the different parts of *S. melongena* (stem, leaves and flowers). From this observation, it appears that of all the pests inventoried, *Arhopalus rusticus* is in the majority with 24.25% followed by *Helicoverpa amigera* with 20.04% and *Aphis gossipii* with 15.92% and preferentially attack the leaves as they are more fragile and contain nutrients and water. This result is similar to those of (Yarou *et al.,* 2017; EPPO, 2017 and Ephytia, 2020) which showed that Solanaceae are attacked mainly by Aphididae, Coreidae, Aleyrodidae. Table 1 presents the pests recorded on the different organs of aubergine.

III.2.2. Pollinators inventoried on *Solanum melongena*

From 25 to 30 October 2020, 70 visits of three insect species were counted on 120 flowers of *S. melongena* left to pollinate freely. Table 2 presents the list of these insects and their percentage of visits. From this table it appears that among the floricultural insects of *S. melongena*, *A. mellifera* occupies the first place with 77.14% of the visits. Catherine (2019) showed that this Apideae (*Apis mellifera*) is known as the main pollinator of this Solanaceae. Furthermore Agnes (2017) observed that bumblebees and xylocopes are more suitable for pollinating aubergine flowers as they are pro-agents of vibratory pollinated flowers.

The high frequency of visits of *A. mellifera* on aubergine flowers could be explained by the good attractiveness of the floral products of this plant to this bee.

Table 1: Insect pests recorded on *Solanum melongena* number and percentage of the number of individuals of these different insects in Bocklé in 2020.

Order	Family	Gender, Species	Bodies						nT	PT (%)
			Stem		Sheet		Flower			
			$n1$	$p1$ (%)	$n2$	$p2$ (%)	$n3$	$p3$ (%)		
Coleoptera	Cerambycidae	*Arhopalus rusticus*	-	-	1748	26,81	-	-	1748	24,25
Hemiptera	Aphididae	*Aphis gossypii*	18	3,65	1130	17,32	-	-	1148	15,92
	Coreidae	*A. femorata*	424	86,00	-	-	100	51,81	524	7,27
	Aleyrodidae	*T. vaporariorum*	31	6,29	700	10,74	-	-	731	10,14
		Dystercus sp.	-	-	18	0,27	-	-	18	0,25
Hymenoptera	Formicidae	*Crematogaster sp.*	20	4,06	632	9,69	-	-	652	9,05
Lepidoptera	Noctuidae	*Helicoverpa armigera*	-	-	1445	22,15	-	-	1445	20,05
	Crambidae	*Haritalodes derogata*	-	-	100	1,54	64	33,16	164	2,27
	Sphingidae	*Agrius sp.*	-	-	400	6,14	29	15,03	429	5,95
Orthoptera	Acrididae	*Chorthippus brunneus*	-	-	18	0,27	-	-	18	0,25
	Tettigoniidae	*T. viridissima*	-	-	325	4,98	-	-	325	4,51
Mantodea	Mantidae	*Mantis religiosa*	-	-	6	0,09	-	-	6	0,08
Total		**12 species**	**493**	**100**	**6522**	**100**	**193**	**100**	**7208**	**100**

n: number of individuals; *PT*: percentage of individuals; *P1* = (*n1* / 493) * 100; *P2* = (*n2* / 6522) * 100; *P3* = (*n3* / 193) * 100; *PT* = (*nT* / 7208) * 100

Table 2: Pollinating insects recorded on *Solanum melongena* flowers, number and percentage of visits of these different insects to Bocklé in 2020.

Order	Family	Genus, Species, Floral product	n	P (%)
Hymenoptera	Apidae	*Apis mellifera* (Po)	**54**	**77,14**
	Apidae	*Xylocopa* (Po)	13	18,57
	Crabronidae	*Crabro sp.* (Po)	3	4,29
Total		**03 species**	**70**	**100 %**

n: number of visits on 120 inflorescences of treatment 1 in 5 days of observation; *P*: percentage of visits = (n / 70) * 100; **Po:** pollen

Figure 24 shows some insect pests (A) and pollinators (B) of *Solanum melongena* in Bocklé in 2020.

Figure 24: Some insect pests **(A)** and pollinators **(B)** of *Solanum melongena* in Bocklé
in October 2020

III.3 Effect of the aqueous extract of the leaves of *Azadirachta indica, Carica papaya* **and** *Senna didymobotrya* **on the population of insect pests**

III.3.1. *Arhopalus rusticus*

Table 3 presents the effect of aqueous powders of *A. indica*, *C. papaya* and *S. didymobotrya* leaves on the population of *A. rusticus*. From this table it can be seen that the aqueous plant powders reduced the population of *Arhopalus rusticus* in the bio-insecticide-treated subplots very significantly ($P < 0.001$) compared to the control and synthetic insecticide-treated subplots (Optimal).

Table 3: Population density of *Arhopalus rusticus* from day 36^e after transplanting

Treatment	Days After Planting (DAP)						$F(9,28)$
	36	43	50	57	64	71	
Witness	$9,75\pm0,85$	$12,25\pm1,79$	$9,75\pm0,85^a$	$10,25\pm2,95$	$12,00\pm1,29$	$9,00\pm0,00$	$13,27^{ns}$
C. papaya	$9,66\pm0,88^{CD}$	$8,33\pm2,33^{CD}$	$11,00\pm0,81^{Cab}$	$16,75\pm0,62^A$	$11,50\pm0,64^C$	$8,75\pm0,25^{BCD}$	22.90^{***}
S. didymobotrya	$10,00\pm1,00^{BC}$	$12,50\pm1,84^{AB}$	$6,50\pm2,39^{BCb}$	$17,00\pm1,08^A$	$12,50\pm1,04^{AB}$	$8,66\pm0,33^{BC}$	9.45^{***}
A. indica	$10,00\pm1,00^A$	$10,66\pm3,84^B$	$14,0\pm1,08^{BCab}$	$19,00\pm1,08^C$	$10,75\pm0,47^B$	$7,50\pm0,64^{BC}$	43.09^{***}
Optimal	$9,00\pm0,57$	$11,66\pm2,60$	$11,0\pm4,02^{ab}$	$10,00\pm1,00$	$10,50\pm0,64$	$8,50\pm0,28$	7.40^{ns}
$F(4, 15)$	$0,20^{ns}$	$0,65^{ns}$	$3,60^*$	$1,33^{ns}$	$0,72^{ns}$	$2,51^{ns}$	

Each value represents the mean ± MSE. Means in the same row and column followed by the same lower and upper case letters are not statistically different according to the Tukey test at the 5% level. ns = non-significant difference ($P > 0.05$); (*$P < 0.05$) significant difference. (***$P < 0.001$) very highly significant difference.

III.3.2. *Helicoverpa armigera*

Table 4 shows the effect of aqueous powders of *A. indica*, *C. papaya* and S. didymobotrya leaves on the H. armigera population. *S. didymobotrya* on the *H. armigera* population. It was found that the aqueous powders of *A. indica* and *S. didymobotrya* and the pesticide Optimal had a very highly significant effect ($P < 0.001$) on the population of *Helicoverpa armigera* compared to the aqueous powders of *C. papaya* which had a highly significant effect ($P < 0.01$) on this insect. On the other hand, the negative control has no significant effect ($P > 0.05$) on the *Helicoverpa armigera* population.

Table 4: Population density of *Helicoverpa armigera* from 36^e days after transplanting

Treatment	Days After Planting (DAP)						$F_{(9,28))}$
	36	43	50	57	64	71	
Witness	$11,25\pm1,65$	$9,00\pm2,67$	$10,00\pm3,53$	$9,00\pm1,08$	$10,50\pm0,64$	$8,00\pm0,40$	$6,18^{ns}$
C. papaya	$9,75\pm1,18^{ABC}$	$11,0\pm1,68^{AB}$	$9,25\pm3,19^{ABC}$	$15,25\pm1,75^A$	$11,00\pm0,91^{AB}$	$7,75\pm0,47^{BC}$	6.07^{**}
S. didymobotrya	$8,00\pm1,47^C$	$9,00\pm1,54^B$	$9,75\pm2,49^B$	$16,50\pm1,32^A$	$12,50\pm1,55^B$	$8,33\pm0,66^C$	6.82^{***}
A. indica	$7,50\pm1,19^B$	$9,50\pm1,50^C$	$8,75\pm2,13^B$	$12,75\pm1,03^A$	$11,25\pm0,85^A$	$7,00\pm0,40^B$	20.4^{***}
Optimal	$8,75\pm0,25^{BC}$	$6,50\pm1,93^{BC}$	$7,75\pm2,95^{BC}$	$13,25\pm1,18^A$	$12,75\pm0,85^{AB}$	$7,00\pm0,00^{BC}$	7.15^{***}
$F_{(4, 15)}$	$1,43^{ns}$	$0,70^{ns}$	$0,11^{ns}$	$1,39^{ns}$	$0,93^{ns}$	$1,93^{ns}$	

Each value represents the mean ± MSE. Means in the same row and column followed by the same lower and upper case letters are not statistically different according to the Tukey test at the 5% level. ns = non-significant difference ($P > 0.05$); (**$P < 0.01$) highly significant difference. (***$P < 0.001$) very highly significant difference.

III.3.3. *Aphis gossipii*

Table 5 shows the effect of aqueous powders of the leaves of *A. indica*, *C. papaya* and *S. didymobotrya* on the population of *Aphis gossipii*. It was found that the aqueous powders of *A. indica* and C. *papaya* caused a very highly significant ($P < 0.001$) decrease in the population of *Aphis gossipii* compared to the aqueous powders of *S. didymobotrya* and the positive control which caused a highly significant ($P < 0.01$) decrease in the population of this insect. In contrast, the control had no significant effect ($P > 0.05$) on the population of *Aphis gossipii*.

Table 5: Population density of *Aphis gossipii* from 36[e] days after transplanting

Treatment	Days After Planting (DAP)						$F_{(9,28)}$
	36	43	50	57	64	71	
Witness	$7,25\pm0,85$	$13,67\pm2,02$	$9,50\pm0,50$	$14,50\pm0,64^{ab}$	$10,75\pm0,85$	$7,75\pm0,47$	$23,34^{ns}$
C. papaya	$9,00\pm0,57^{B}$	$12,33\pm1,66^{AB}$	$10,00\pm1,52^{B}$	$15,75\pm0,^{62Aa}$	$11,50\pm0,64^{B}$	$8,33\pm0,33^{B}$	$28,01^{***}$
S. didymobotrya	$10,00\pm1,00^{B}$	$10,00\pm0,57^{B}$	$14,00\pm5,00^{A}$	$14.00\pm0.^{70Aabc}$	$11,50\pm0,64^{AB}$	$8,00\pm0,00^{BC}$	$10,63^{**}$
A. indica	$9,50\pm0,50^{A}$	$9,00\pm0,57^{B}$	$7,33\pm3,17^{B}$	$10,75\pm1,18^{Ac}$	$10,75\pm0,47^{A}$	$8,00\pm1,00^{B}$	$25,36^{***}$
Optimal	$8,66\pm0,66^{BC}$	$9,33\pm0,88^{B}$	$9,50\pm0,50^{B}$	$11,25\pm1,10^{Abc}$	$11,75\pm0,85^{A}$	$7,00\pm0,00^{BC}$	$18,66^{**}$
$F_{(4, 15)}$	$1,88^{ns}$	$2,48^{ns}$	$0,75^{ns}$	$5,91^{**}$	$0,43^{ns}$	$0,48^{ns}$	

Each value represents the mean ± MSE. Means in the same row and column followed by the same lower and upper case letters are not statistically different according to the Tukey test at the 5% level. ns = non-significant difference ($P > 0.05$); (**P < 0.01) highly significant difference. (***P < 0.001) very highly significant difference.

III.3.4. *Trialeurodes vaporariorum*

Table 6 shows the effect of aqueous powders of *A. indica*, *C. papaya* and S. didymobotrya leaves on the T. vaporariorum population. *S. didymobotrya* on the *T. vaporariorum* population. It follows from this table that the aqueous powders of *A. indica* and the positive control have a very highly significant effect ($P < 0.001$) on the T. vaporariorum population, while the aqueous powders of C. *papaya* and S. *didymobotrya* have a highly significant effect ($P < 0.01$) on this species. On the other hand, the negative control had no significant effect ($P > 0.05$) on T. *vaporariorum*.

Table 6: Population density of *Trialeurodes vaporariorum* from 36[e] days after transplanting

Treatment	Days After Planting (DAP)						$F_{(9,28)}$
	36	43	50	57	64	71	
Witness	$2,50\pm0,50$	$8,00\pm0,00$	$4,66\pm2,02$	$5,25\pm1,31^{b}$	$12,25\pm0,85$	$7,50\pm0,86$	$5,77^{ns}$
C. papaya	$2,50\pm0,50^{C}$	$8,00\pm0,00^{BC}$	$6,00\pm1,00^{BC}$	$5,00\pm1,08^{BCb}$	$11,75\pm0,85^{A}$	$6,50\pm1,89^{BC}$	$11,49^{**}$
S. didymobotrya	$3,00\pm0,00^{B}$	$8,00\pm0,00^{BC}$	$7,00\pm2,00^{BC}$	$10,00\pm1,54^{Ba}$	$9,75\pm0,47^{B}$	$8,25\pm0,47^{B}$	$84,37^{**}$
A. indica	$3,00\pm0,00^{C}$	$8,00\pm0,00^{B}$	$6,66\pm1,76^{B}$	$8,25\pm0,95^{Bab}$	$9,75\pm0,47^{A}$	$7,33\pm0,33^{B}$	$84,37^{***}$
Optimal	$4,00\pm0,00^{C}$	$4,50\pm3,50^{C}$	$6,66\pm1,20^{C}$	$8,75\pm0,25^{Bab}$	$11,50\pm0,86^{A}$	$9,00\pm0,57^{B}$	$8,02^{***}$
$F_{(4, 15)}$	$1,87^{ns}$	$0,18^{ns}$	$1,21^{ns}$	$4,99^{**}$	$0,44^{ns}$	$1,03^{ns}$	

Each value represents the mean ± MSE. Means in the same row and column followed by the same lower and upper case letters are not statistically different according to the Tukey test at the 5% level. ns = non-significant difference ($P > 0.05$); (**P < 0.01) highly significant difference. (***P < 0.001) very highly significant difference.

III.3.5. *Crematogaster sp.*

Table 7 presents the effect of aqueous powders of *A. indica*, *C. papaya* and *S. didymobotrya* leaves on the population of *Crematogaster sp. indica* had a very highly significant effect ($P < 0.001$) on *Crematogaster sp.* compared to *C. papaya, S. didymobotrya* and the synthetic insecticide which had a highly significant effect ($P < 0.01$) on the latter population. The control had no significant effect ($P > 0.05$) on *Crematogaster sp.*

Table 7: Population density of *Crematogaster sp.* from 36^e days after transplanting

Treatment	Days After Planting (DAP)						$F_{(9,28)}$
	36	43	50	57	64	71	
Witness	$5,25\pm0,85$	$9,00\pm1,15$	$3,33\pm1,33$	$9,25\pm0,85^b$	$2,00\pm1,00$	$4,00\pm3,00$	9.30^{ns}
C. papaya	$3,00\pm2,00^C$	$9,50\pm1,50^{BC}$	$6,33\pm2,66^C$	$12,75\pm0,47^{Aa}$	$0,00\pm0,00^C$	$8,00\pm0,00^{BC}$	$4.51**$
S. didymobotrya	$1,50\pm0,50^B$	$9,00\pm8,00^B$	$4,33\pm2,84^B$	$11,00\pm0,40^{Ab}$	$0,00\pm0,00^B$	$8,50\pm0,50^B$	$7.94**$
A. indica	$3,00\pm1,08^C$	$6,33\pm1,33^B$	$1,66\pm0,33^C$	$10,00\pm0,40^{Ab}$	$0,00\pm0,00^C$	$8,00\pm0,00^B$	$15.98***$
Optimal	$2,50\pm0,64^B$	$7,33\pm3,17$	$1,33\pm0,33^B$	$12,75\pm1,01^{Aa}$	$0,00\pm0,00^B$	$7,00\pm0,00^B$	$5.01**$
$F_{(4, 15)}$	$0,92^{ns}$	$0,35^{ns}$	$0,31^{ns}$	$5,43**$	$2,57^{ns}$	$0,71^{ns}$	

Each value represents the mean $\pm$ MSE. Means in the same row and column followed by the same lower and upper case letters are not statistically different according to the Tukey test at the 5% level. ns = non-significant difference ($P > 0.05$); (**P < 0.01) highly significant difference. (***P < 0.001) very highly significant difference.

III.3.6. *Acanthocephala femorata*

Table 8 shows the effect of aqueous powders of *A. indica*, *C. papaya* and S. didymobotrya leaves on the A. femorata population. S. didymobotrya on the *A. femorata* population. It was found that the A. *indica* and *C. papaya* aqueous powders had a very highly significant effect ($P < 0.001$) on the population of this insect compared to the negative and positive control subplots which showed no significant effect ($P > 0.05$). Aqueous powders of *S. didymobotrya* had a highly significant effect ($P < 0.01$) on the population of the latter.

Table 8: Population density of *Acanthocephala femorata* from day 36^e after transplanting

Treatment	Days After Planting (DAP)						$F_{(9,28)}$
	36	43	50	57	64	71	
Witness	$3,00\pm0,00$	$8,00\pm0,00$	$5,00\pm0,00$	$7,00\pm0,00$	$7,25\pm2,42$	$8,50\pm0,50$	64.78^{ns}
C. papaya	$6,00\pm0,00$	$7,00\pm0,00$	$7,00\pm0,00$	$0,00\pm0,00$	$10,25\pm0,75$	$9,00\pm0,00$	$34.87***$
S. didymobotrya	$7,00\pm0,00^B$	$7,00\pm0,00^B$	$10,00\pm0,00$	$8,00\pm1,00^B$	$12,00\pm0,91^B$	$8,33\pm0,33^B$	$12.87**$
A. indica	$6,00\pm0,00$	$0,00\pm0,00$	$9,00\pm0,00^B$	$5,50\pm0,50$	$11,25\pm0,75$	$8,00\pm1,00$	$17.61***$
Optimal	$3,00\pm0,00$	$4,50\pm3,50$	$7,00\pm0,00$	$8,00\pm0,00$	$11,00\pm0,40$	$7,00\pm0,00$	3.73^{ns}
$F_{(4, 15)}$	-	$0,14^{ns}$	-	$3,00^{ns}$	$2,11^{ns}$	$2,80^{ns}$	

Each value represents the mean $\pm$ MSE. Means in the same row and column followed by the same lower and upper case letters are not statistically different according to Tukey's test at the 5% level. ns = non-significant difference ($P > 0.05$); (**P < 0.01) highly significant difference; (***P < 0.001) very highly significant difference; (-) estimation of the F-value is not possible due to equal variance

It was generally noted that all bio-insecticides have a negative impact on the pest population thus allowing their control. *Azadirachta indica was* very effective in reducing the populations of Aphididae, Noctuidae and Coreidae in a highly significant manner ($P < 0.001$). These results are similar to those obtained by Mondédji (2017); Sane (2018) who also showed that aqueous extracts of neem reduced the populations of Aphididae. As for *C. papaya* and *Senna didymobotrya*, they highly significantly reduced the population of *Crematogaster sp.* and *Trialeurodes vaporariorum.*

The decrease in insect pest density in subplots treated with aqueous powder extracts could be a consequence of various biological effects that would effectively reduce the insects in the larval stage before they become adults, as well as the reproduction rate (Cressman, 2021), and also to variations in climatic conditions.

It should be noted, however, that insecticidal plants do not eliminate all insect pests (Sylvie & Martin, 2017). Insects exposed to treatments may take several days to die, or they may simply move away. Insecticidal plants can be directly toxic, but can also often act by repulsion, preventing feeding, growth regulation or egg laying (Anjarwalla *et al.,* 2016).

The results show that botanical insecticides could effectively control the pest population. Studies by Sane (2018) and Aribi (2020) showed that *A. indica* contains major compounds such as azadirachtin and tetranortriterpenoids that could be responsible for this reduction in insect pest populations. Tahiri (2010) showed in Côte d'Ivoire that papaya would possess terpenoids and tannins that have insecticidal, repellent and toxic effects on insects. Cheikh (2015) showed in Senegal that senna would have the hetere which would control *Caryedon serratus,* an insect pest of groundnuts.

III.4 Effect of aqueous extract of *Azadirachta indica, Carica papaya* **and Senna didymobotrya leaf powders on physiological parameters** *Senna didymobotrya* **on physiological parameters.**

III.4.1. Number of leaves per plant

Figure 25 shows the effect of the aqueous extract of the leaf powders of *A. indica, C. papaya* and *S. didymobotrya* on the number of leaves per plant. The figure shows that there is a very highly significant difference ($P < 0.001$). The number of leaves per plant was higher in the subplots treated with *A. indica* than in all other subplots. We observed an overall increase in all subplots between 43[e] and 50[e] DAR with a peak in the number of leaves observed on 64[e] DAR on the subplot treated with *A. indica.* We also observed an overall decrease in the number of leaves in all treated and untreated subplots from 65e JAR.

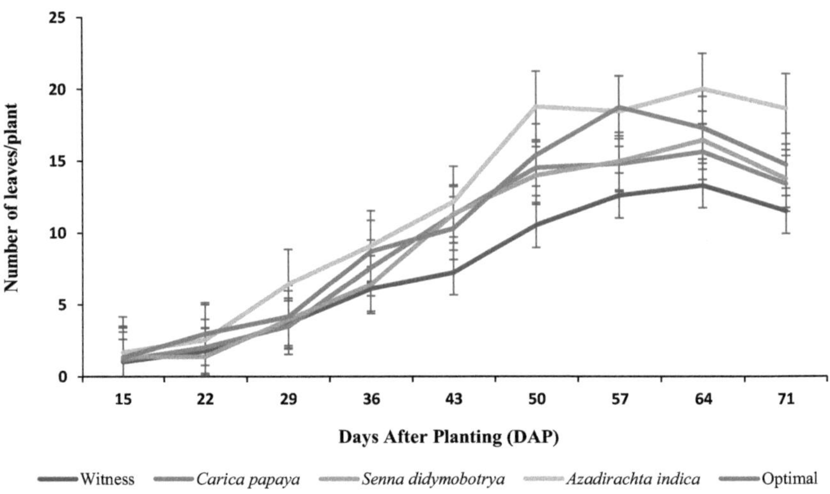

Figure 25: Dynamics of the number of leaves under different insecticide treatments in Bocklé in 2020

III.4.2. Number of flowers per plant

The results of the effect of botanical insecticide treatments on the number of flowers per plant showed that there was no significant difference ($P > 0.05$) between treatments (Table 9). This could be explained by the low presence of flower pests in the field at our experimental site.

Table 9: Flower density per plant of each treatment from day 50^e after transplanting

Treatments	Days After Transplanting (JAR)				$F_{(3,12)}$
	50	57	64	71	
Witness	$1,36 \pm 0,30$	$2,28 \pm 0,29$	$2,12 \pm 0,39$	$1,86 \pm 0,29$	$1,44^{ns}$
C. papaya	$2,18 \pm 0,51$	$2,32 \pm 0,31$	$2,26 \pm 0,06$	$1,98 \pm 0,25$	$0,21^{ns}$
S. didymobotrya	$2,00 \pm 0,37$	$2,16 \pm 0,11$	$2,20 \pm 0,09$	$2,08 \pm 0,15$	$0,14^{ns}$
A.indica	$2,15 \pm 0,50$	$2,22 \pm 0,33$	$2,36 \pm 0,13$	$1,97 \pm 0,25$	$0,24^{ns}$
Optimal	$2,18 \pm 0,41$	$2,37 \pm 0,21$	$2,55 \pm 0,20$	$2,14 \pm 0,19$	$0,40^{ns}$
$F_{(4, 15)}$	$0,65^{ns}$	$0,70^{ns}$	$0.60ns$	$0.20ns$	

Each value represents the mean $\pm$ MSE. ns = non-significant difference ($P > 0.05$).

III.4.3. Number of fruits per plant

Figure 26 shows the dynamics of the number of fruits per plant in the different subplots. From this figure it can be seen that all subplots have a high fruit rate at 64 JAR with the highest peak in the subplots treated with *A. indica* at 64^e JAR. Between 64^e and 71^e JAR, there was a slight decrease in the average number of fruits in the negative control and *S. didymobotrya* treated subplots and a slight increase in the number of fruits in the A. indica treated subplots.

of the average number of fruits in the positive control and *A. indica* and *C. papaya* treated subplots.

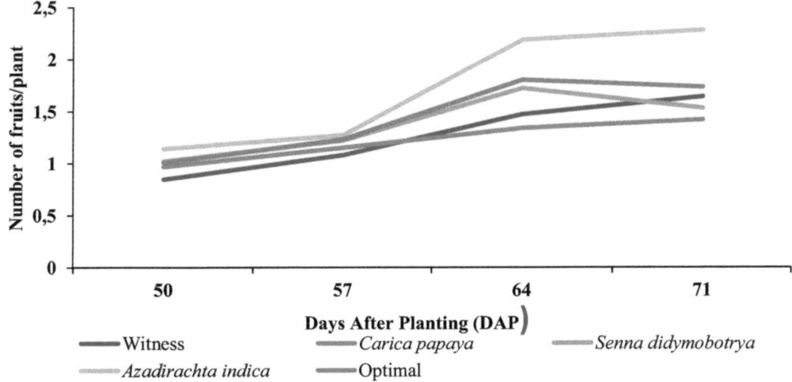

Figure 26: Dynamics of the number of fruits under the effect of different insecticide treatments from 50 DAR in Bocklé in 2020

III.5. Activity of *Apis mellifera* on the flowers of *Solanum melongena*

III.5.1. Collected floral products

At the flowers of *S. melongena*, *A. mellifera* collected pollen exclusively and intensively. Out of 54 recorded visits of *Apis mellifera*, 54 (100%) were for pollen collection.

Figure 27: *Apis mellifera* worker collecting pollen from a *Solanum melongena* flower in Bocklé, October 2020

III.5.2. Rhythm of visits of *Apis mellifera* according to the daily time slots of observation and under the effect of the insecticide treatments

Figure 28 shows the distribution of *A. mellifera* visits in the different treated and untreated sub-plots on *S. melongena* flowers according to the daily observation time slots in Bocklé in 2020. It appears from this figure that *A. mellifera* has a period of activity between 8 and 16 h to visit the flowers of *S. melongena* with a maximum activity between 10 and 11 h ($X^2 = 0.05$; *ddl* =

1; $P < 0.05$) in all the sub-plots except those treated with *S. didymobotrya*. On the other hand, the subplots treated with *S. didymobotrya* had a maximum activity between 12 and 13 h. According to Langlois & Alban (2019), these periods of maximum activity would correspond to the availability of the floral product of this species (pollen). *On the* other hand, the low peak and the shift of the peak activity of the subplots treated with *S. didymobotrya* from the rest of the subplots could be explained by the fact that the odour of the aqueous extract of *S. didymobotrya* is known to have repulsive effects on bees (Tabuti, 2016).

The low activity of *A. mellifera* observed at the flowers during the other time slots would be related to the decrease of floral products of this Solanaceae.

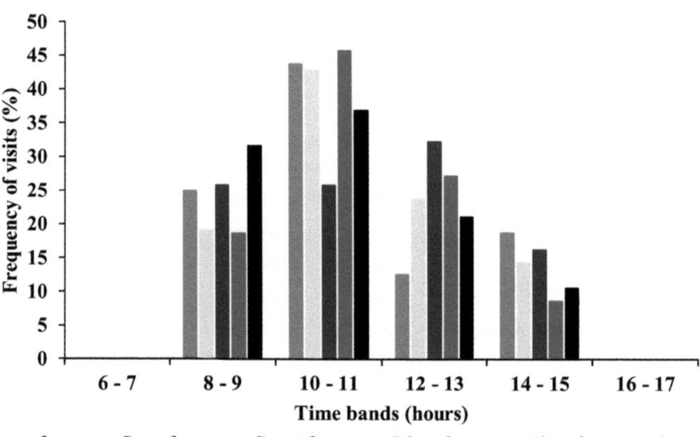

St: no treatment; Cp: *Carica papaya*; Sd: *Senna didymobotrya*; Ai: *Azadirachta indica*; Ic: chemical insecticide.

Figure 28: Distribution of *Apis mellifera* visits on *Solanum melongena* flowers in the different treated sub-plots according to daily time slots in Bocklé in 2020

III.5.3. Abundance of pests and foragers

Figure 29 shows the abundance of the major insect pests of *S. melongena*. From this figure, it appears that aubergine is more attacked by the Coleoptera population, especially *Arhopalus rusticus*.

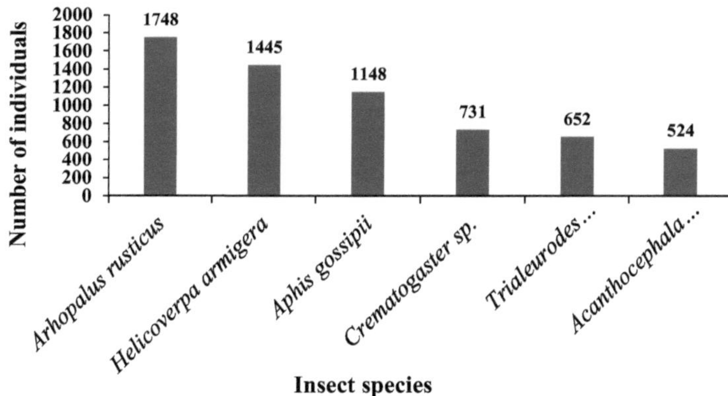

Figure 29: Abundance of insect pests of *Solanum melongena* at Bocklé in 2020

The highest number of *Apis mellifera* workers active on a *S. melongena* flower in general is 1 ($n = 443$; $s = 0$). This low density would be due to the small landing area of the flowers of this Solanaceae.

The average abundance per 1000 flowers is 14 ($n = 443$; $s = 1.78$).

The average abundance per 1000 flowers shows the good attractiveness of *A. mellifera* pollen to *S. melongena*. This is related to their ability to recruit large numbers of foragers to exploit an attractive food source (Louveaux, 1984; Seeley *et al.* , 1991). Table 10 shows the abundance of foragers in the different subplots.

Table 10: Abundance of foragers in the different subplots

Treatments	Flower		Per 1000 flowers	
	n	*m* ±es	*n*	*m* ±es
Witness	87	$1,00 \pm 0,00$	87	75.44 ± 4.84 [bc]
Carica papaya	112	$1,00 \pm 0,00$	112	72.54 ± 3.71 [abc]
Senna didymobotrya	74	$1,00 \pm 0,00$	74	56.84 ± 3.96 [c]
Azadirachta indica	82	$1,00 \pm 0,00$	82	62.41 ± 3.69 [ab]
Optimal	88	$1,00 \pm 0,00$	88	59.22 ± 3.22 [a]

*F(4, 438) - F(4, 438) 4.33***

Each value represents the mean ± MSE. Means in the same row and column followed by the same lower and upper case letters are not statistically different according to Tukey's test at the 5% level. (**P < 0.01) highly significant difference; (-) estimation of F-value is not possible due to equal variance.

Abundance per 1000 flowers: Overall: $F = 4.33$ ($ddl1 = 4$; $ddl2 = 438$; $P > 0.01$; **HS**). **Pairwise comparison:** Tem / Cp: $t = 0.82$ ($ddl = 197$; $P >0.05$; **NS**); Tem / Sd: $t = 4.01$ ($ddl = 159$; $P < 0.001$; **HRT**); Tem / Ai: $t = 2.90$ ($ddl = 167$; $P > 0.001$; **HS**); Tem / Op: $t = 3.79$ ($ddl = 173$; $P > 0.001$;

HS); Cp / Sd: $t = 3.94$ (*ddl* $= 184$; $P > 0.001$; **HS**); Cp / Ai: $t = 2.62$ (*ddl* $= 192$; $P < 0.001$; **THS**); Cp / Op: $t = 3.64$ (*ddl* $= 198$; $P > 0.001$; **HS**); Sd / Ai: $t = 1.39$ (*ddl* $= 154$; $P > 0.05$; **NS**); Sd / Op: $t = 0.63$ (*ddl* $= 160$; $P > 0.05$; **NS**); Ai / Op: $t = 0.88$ (*ddl* $= 168$; $P > 0.05$; **NS**). Cp: *Carica papaya,* Sd : *Senna didymobotrya,* Az: *Azadirachta indica,* Ic: Optimal 20 sp, St: Control.

- Overall, there was a highly significant difference between the means of forager abundance per 1000 flowers ($F = 4.33$; *ddl1* $= 4$; *ddl2* $= 438$; $P < 0.01$) between the different treatments. -

The mean abundance per 1000 flowers was 73 foragers in the subplots treated with *C. papaya* and 59 foragers in the subplots treated with Optimal 20 SP. The difference between these two means was highly significant ($t = 3.64$; *ddl* $= 198$; $P < 0.01$).

The difference between the average abundance of foragers per 1000 flowers in the *C. papaya* treated and control subplots was not significant. However, this difference is very highly significant, and highly significant between the subplots treated with *S. didymobotrya* and *A. indica*. This difference is also highly significant in the subplots treated with Optimal. The low abundance of foragers in the subplots treated with *S. didymobotrya* and Optimal could be explained by the repellent effect of these synthetic and botanical insecticides on these floricultural insects (Son *et al.*, 2017; Nabie, 2018; Soro, 2019). The higher abundance per 1000 flowers observed in the subplots treated with *C. papaya* highlights the good attractiveness of *S. melongena* pollen and the influence of the odoriferous elements contained in the aqueous extracts of this plant towards *A. mellifera* workers.

III.5.4. Rhythm of the visits of *Apis mellifera* according to the rhythm of blooming of the flowers of *Solanum melongena*

Figure 30 shows the number of open flowers of *S. melongena* and the number of visits of A. mellifera in each sub-plot according to the observation dates. visits of *A. mellifera* in each sub-plot according to the observation dates.

From this figure it can be seen that overall the number of visits of *A. mellifera* is higher on a plant when the number of opened flowers on it is higher and it is even higher when there is a complementary and/or additional supply of plant pheromones due to the effect of insecticide treatments. This is shown in our figure by a maximum number of flowers opened and peak visits recorded on 29/10/20. This result is consistent with that of Faegri & Pijl (1979) who indicated that the number of flowers open is an essential factor that plays an important role in the orientation of insects towards the flowers of *S. melongena*. As a result, there is a high correspondence between the percentage of flowering and the number of pollinating insects (Ferland, 2014).

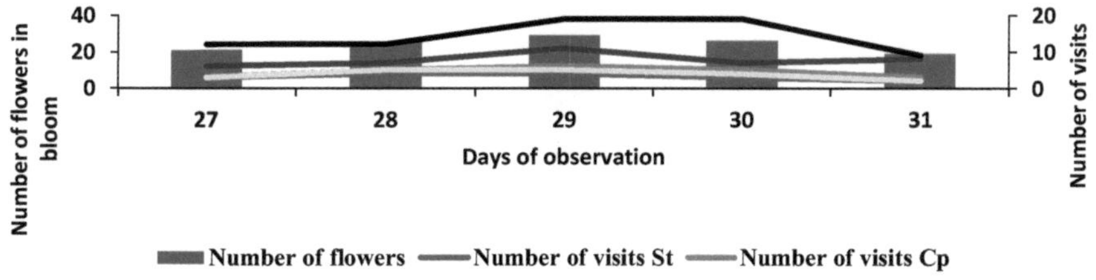

St: no treatment; Cp: *Carica papaya*; Sd: *Senna didymobotrya*; Ai: *Azadirachta indica*; Ic: chemical insecticide.

Figure 30: Visitation rate of *Apis mellifera in* relation to the flowers of *Solanum melongena* in Bocklé in 2020

Overall, we have a positive and highly significant correlation between the number of flowers opened and the number of visits of *Apis mellifera* in :
- Sub-plots treated with *C. papaya* (r = 0.83; *ddl* = 4; P < 0.01);
- Sub-plots treated with *S. didymobotrya* (r = 0.17; *ddl* = 4; P < 0.01);
- Sub-plots treated with *A. indica* (r = - ; *ddl* = ; P -) ;
- Control subplots (r = 0.38 ; *ddl* = 4 ; P < 0.01) ;
- Subplots treated with Optimal 20 SP (r = 0.20 ; *ddl* = 8 ; P < 0.01).

III.5.5. Duration of visits by flower

In general, the average duration of an *A. mellifera* visit per *S. melongena* flower was 8.01 sec (n = 722; s = 0.12) in all subplots. Table 11 shows the duration of visits per flower of *A. mellifera* in the different treated subplots.

Table 11: Duration of visits per *Apis mellifera* flower in the different treated sub-plots

Treatments	Pollen	
	n	**$m \pm$ es**
Witness	124	$8,11 \pm 0,25^{ab}$
Carica papaya	102	$8,70 \pm 0,22^{a}$
Senna didymobotrya	146	$7,17 \pm 0,40^{a}$
Azadirachta indica	206	$8,72 \pm 0,20^{b}$
Optimal	144	$7,71 \pm 0,30^{ab}$
F(4, 717)		4.78^{**}

Each value represents the mean ± MSE. Means in the same column followed by the same lower case letters are not statistically different according to the Tukey test at the 5% level. (** P < 0.01) highly significant difference. Cp: *Carica papaya*, Sd: *Senna didymobotrya*, Az: *Azadirachta indica*, Op: Optimal, Tem: control.

Comparison of visit times of *Apis* **mellifera/flower**: Overall: F = 4.78 (ddl1 = 4; ddl2 = 717; P < 0.01; **HS**).

Pollen collection: Overall: *F* = 4.78 (*ddl1* = 4; *ddl2* = 717; *P* < 0.01; HS).Pairwise comparison: Tem / Cp: *t* = 1.73 (*ddl* = 224; *P* > 0.05; **NS**); Tem / Sd: *t* = 2.32 (*ddl* = 268; *P* < 0.05; **S**); Tem / Ai: *t* =1.08 (*ddl* = 328; *P* > 0.05; **NS**); Tem / Op: *t* = 1.00 (*ddl* = 266; *P* > 0.05; **NS**); Cp / Sd: *t* = 3.31 (*ddl* = 246; *P* > 0.001; **HS**); Cp / Ai: *t* = 1.64 (*ddl* = 306; *P* > 0.05; **NS**); Cp / Op: *t* = 1.97 (*ddl* = 244; *P* < 0.05; **S**); Sd / Ai: *t* = 3.78 (*ddl* = 350; *P* > 0.001; **HS**); Sd / Op: *t* = 2.14 (*ddl* = 288; *P* > 0.05; **NS**); Ai / Op: *t* = 1.97 (*ddl* = 248; *P* < 0.05; **S**)

- Overall, the difference is highly significant for pollen collection (*F* = 4.78; *ddl1* = 4; *ddl2* = 717; *P* < 0.01; ***HS***).

- The visit duration for pollen collection was 8.72 sec (*n* = 206; *s* = 0.20) in subplots treated with *A. indica* and 7.17 sec (*n* = 146; *s* = 0.40) in subplots treated with *S. didymobotrya*. The difference between these two means is highly significant (*t* = 3.78; *ddl* = 350; *P* < 0.01; ***HS***).

- The visit duration for pollen collection was 8.72 sec (*n* = 206; *s* = 0.20) in the *A. indica* treated subplots and 7.71 sec (*n* = 144; *s* = 0.30) in the Optimal treated subplots. The difference between these two means is significant (*t* = 1.97; *ddl* = 248; *P* < 0.05; ***S***).

- The visit duration for pollen collection was 8.70 sec (*n* = 102; *s* = 0.22) in the *C. papaya* treated

subplots and 7.71 sec (*n* = 144; *s* = 0.30) in the Optimal treated subplots. The difference between the two means is significant (*t = 1.97; ddl = 244; P < 0.05; S*).

In general, for all floral products searched, the duration of visits varied with the availability of pollen and the different insecticide treatments and the positive control. In addition, bees were more attracted to the *A. indica* and *C. papaya* treated sub-plots than to other sub-plots. This could be explained by the high availability and quality of the floral products (Free, 1993; Pierre, 2003); and by the presence of odoriferous elements in the aqueous extracts of these two plants which have an attractive role for bees.

III.5.6. Foraging ethology

On a flower of *S. melongena*, the workers of *A. mellifera* selectively visit the opened flowers. When collecting pollen, they land directly on the anthers and then scrape them with their forelegs and mandibles to expel the pollen, which is then collected and accumulated in the baskets of the hind legs.

The passage from one flower to another is done either by walking or by flying.

The average foraging speed is about 11 flowers per minute ($n = 423$; $s = 0.36$).

Table 12 shows the average variation in foraging by *A. mellifera between* the different treatments.

Table 12: Average variation in foraging speed of *Apis mellifera* according to the different treatments

Insecticides	Witness	*C. papaya*	*S. didybomotrya*	*A. indica*	Optimal	$F_{(4, 424)}$
N	93	117	58	81	80	
Speed (af/min)	$10,27 \pm 0,88$	$9,51 \pm 0,29$	$11,80 \pm 1,47$	$11,08 \pm 0,53$	$11,99 \pm 1,07$	$1,73^{ns}$

Each value represents the mean $\pm$ MSE. ns = non-significant difference ($P > 0.05$).

Comparison of mean foraging speeds of *Apis mellifera* **, af / min:** Overall: $F = 1.73$ (ddl1 = 4; ddl2 = 424; $P > 0.05$; **NS**. 1.73 (*ddl1* = 4; *ddl2* = 424; $P > 0.05$; **NS**).pairwise comparison: Tem / Cp: $t = 0.99$ (*ddl* = 207; $P > 0.05$; **NS**); Tem / Sd: $t = 1.16$ (*ddl* = 148; $P > 0.05$; **NS**); Tem / Ai: $t = 0.91$ (*ddl* = 171; $P > 0.05$; **NS**); Tem / Op: $t = 1.52$ (*ddl* = 170; $P > 0.05$; **NS**); Cp / Sd: $t = 2.19$ (*ddl* = 173; $P < 0.05$; **S**); Cp / Ai: $t = 2.84$ (*ddl* = 196; $P < 0.01$; **HS**); Cp / Op: $t = 2.86$ (*ddl* = 195; $P < 0.01$; **HS**); Sd / Ai: $t = 0.59$ (*ddl* = 137; $P > 0.05$; **NS**); Sd / Op: $t = 0.13$ (*ddl* = 136; $P > 0.05$; **NS**); Ai / Op: $t = 0.90$ (*ddl* = 159; $P > 0.05$; **NS**). Cp: *Carica papaya,* Sd: *Senna didymobotrya,* Ai : *Azadirachta indica,* Ic: Optimal 20 sp, St : Control.

- Overall, the differences between the mean foraging speeds are non-significant ($F = 1.73$; *ddl1* = 4; *ddl2* = 424; $P < 0.05$; *NS*) between the different treatments.

- The average foraging visit was 9.51 flowers per min ($n = 117$; $s = 0.29$) in the subplots treated with *C. papaya* and 11.80 flowers per min ($n = 58$; $s = 1.47$) in the subplots treated with *S. didymobotrya.* The difference between these two means is significant ($t = 2.19$; *ddl* = 173; $P < 0.05$; *S*).

- The average foraging visit was 9.51 flowers per min ($n = 117$; $s = 0.29$) in the subplots treated with *C. papaya* and 11.99 flowers per min ($n = 80$; $s = 1.07$) in the subplots treated with Optimal. The difference between the two means is highly significant ($t = 2.86$; *ddl* = 195; $P < 0.01$; *HS*).

In summary, the average foraging visit varied between the different insecticide treatments but the difference between these average speeds was not significant. In addition, bees had higher speeds in the *S. didymobotrya* and Optimal treated subplots than in the other subplots. The observed consistency in foraging speeds could be justified by the same level of accessibility, the same degree of availability of floral products, the distances between the flowers exploited during the different foraging trips.

III.5.7. Foraging ecology
a. Influence of wildlife

During foraging activity, A. *mellifera,* including other pollinating insects, were disturbed by other insects competing for pollen. Interruptions of visits were caused by collisions between visitors or by approaching a flower already occupied by another visitor. Thus, out of 50 visits studied, 9 (18%) were interrupted by other individuals of *A. mellifera.*

These disturbances resulted in the reduction of the duration of some visits; this forced foragers to visit more flowers during a foraging trip to obtain their optimal pollen load (Tchuenguem *et al.* , 2001).

b. Influence of the surrounding flora

During the flowering period of *S. melongena,* other plants in the vicinity of the study site were visited by *A. mellifera* workers for nectar or pollen. Table 13 shows some of these plants and the intensity of harvesting of each of these floral products. Out of 50 visits of this bee to aubergine flowers, we noted 23 (46%) passages of

flowers of *S. melongena* to those of *Abelmoschus esculentus* and 19 (38%) from *Abelmoschus esculentus* to *S. melongena.*

| | Food taken | |
Plant species	Nectar	Pollen
Abelmoschus esculentus	+++	++
Allium cepa	-	++
Corchorus olitorus	++	++
Hibiscus rabdariffa	++	++
Medicago anguina	++	++
Psidium guajava	-	++
Physalis minima	++	++

| *Vigna unguiculata* | ++ | ++ |
| *Zea mays* | - | ++ |

Table 13: Flower products collected by *Apis mellifera* from the flowers of some neighbouring *Solanum melongena* plants in Bocklé in 2020

- : no harvest; + : very low harvest; ++ : high harvest; +++ : very high harvest

Figure 31 shows the workers of *A. mellifera* collecting pollen from the flowers of some of the neighbouring plant species of *S. melongena* in Bocklé in 2020.

Apis mellifera collecting nectar from the flower of *Abelmoschus esculentus*

Apis mellifera collecting pollen from a *Zea mays* panicle

Apis mellifera collecting pollen from the flower of *Physalis minima*

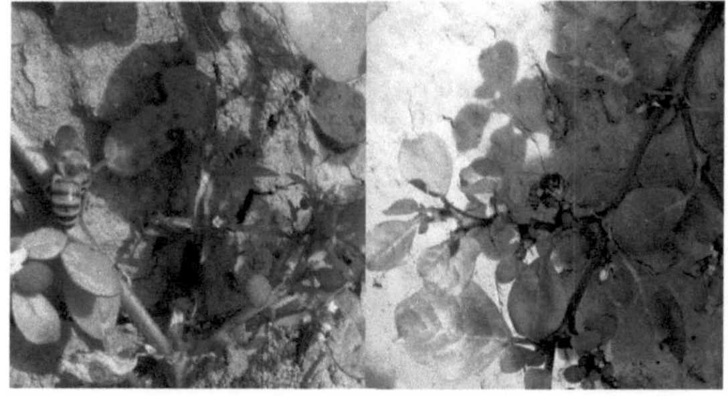

Apis mellifera collecting nectar from the flower of *Medicago anguina*

Apis mellifera collecting pollen from the flower of *Medicago anguina*

Apis mellifera collecting nectar from the flower of *Vigna unguiculata*

Figure 31: *Apis mellifera* workers active on the flowers of five plant species near *Solanum melongena* in Bocklé in October 2020

A. mellifera collected both nectar and pollen from the flowers of *Physalis minima* and *Abelmoschus esculentus*.

Thus, workers of *A. mellifera* were generally faithful to the flowers of *S. melongena* during foraging. They maintain this fidelity to the species during foraging trips; this phenomenon, well known in honey bees, is known as floral constancy (Louveaux, 1984).
According to Greggers & Menzel (1992), a bee usually forages for only one plant species per foraging trip.

c. Influence of some climatic factors

During the observation period, some climatic factors influenced the activity of *A mellifera to a* greater or lesser extent. Out of 50 visits, 16 (32%) were interrupted by strong winds.

During the whole flowering period of this Solanaceae, for the 14 days of observation, two were sunny, five were partly sunny and overcast, three were rainy and then moderately sunny and four were foggy.

These observations are similar to those of Louveaux & Pesson (1984) who underlined that foragers prefer sunny and warm days but limit their activity during rainy days or strong winds. Figure 32 shows the variations of temperature and hygrometry of the study station according to the number of visits of *A. mellifera* on the flowers of *S. melongena* according to the time slots of investigation. It follows that there is no correlation between temperature and the number of visits of *A. mellifera on the one hand* ($r = - 0.20$; $ddl = 4$; $P > 0.05$); and between hygrometry and this same number of visits on the other hand ($r = - 0.55$; $ddl = 4$; $P > 0.05$).

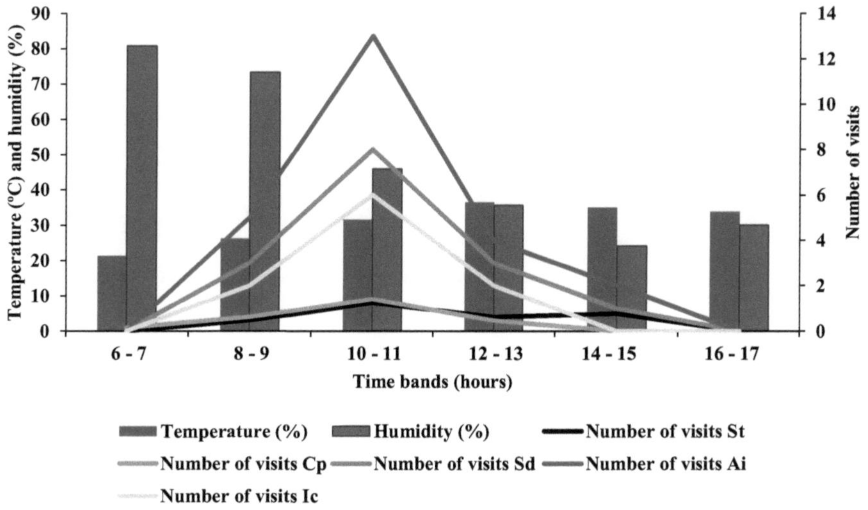

St: no treatment; Cp: *Carica papaya*; Sd: *Senna didymobotrya*; Ai: *Azadirachta indica*; Ic: chemical insecticide.

Figure 32: Daily variation of temperature, relative air humidity and number of visits of *Apis mellifera on Solanum melongena* flowers in Bocklé in 2020

III.6. Beekeeping value of *Solanum melongena*

During the flowering periods of *S. melongena*, there was a good pollen collection activity by the workers of *A. mellifera on the* flowers of this Solanaceae. This result highlights the good attractiveness of the workers of *A. mellifera* towards the pollen of *S. melongena,* and consequently these data obtained allow us to classify this plant species among the pollinating beekeeping plants. Therefore, this Solanaceae can be cultivated and preserved in the apiary environment to increase pollen production as a hive product and to stabilise *A. mellifera* colonies.

III.7. Impact of *Apis mellifera* **on the pollination of** *Solanum melongena*

When collecting pollen from a flower of *S. melongena*, workers of *A. mellifera* were always in contact with the anthers (100%) and the stigma (100%). They could therefore intervene directly in self-pollination by putting the pollen of a flower on its stigma. Therefore, *A. mellifera* increases the pollination possibilities of *S. melongena*.

III.8. Yields of *Solanum melongena* according to the different treatments

Table 14 shows the fruiting rate, average number of seeds and percentage of normal seeds in the different *S. melongena* treatments.

Table 14: Fruiting rate, average number of seeds and percentage of normal seeds in the different treatments of *Solanum melongena*

Each value represents the mean ± MSE. Means of the same row and column followed by the same lower and upper case letters are not statistically different according to the Tukey test at the 5% level. ns = non-significant difference ($P > 0.05$); (*P < 0.05) significant difference; (*P < 0.01) highly significant difference; (*** $P < 0.001$) very highly significant difference. (-) estimation of F-value is not possible due to equal variance; **TF:** fruiting rate; **%GN:** percentage of normal seeds; **FL:** free flowers; **FID:** flowers protected from insects; **FvA:** flowers visited exclusively by *Amegilla* sp. 1; **Fpsv:** flowers: flowers protected, then destined to open and close without visit of insect or any other organism.

Comparison of fruiting rates of the different subplots :

Untreated subplots: T6 / T7: $x2 = 5.78$ (*ddl* = 1; $P < 0.05$; **S**); T6 / T8: $x2 = 0.05$ (*ddl* = 1; $P > 0.05$; **NS**); T6 / T9: $x2 = 0.48$ (*ddl* = 1; $P > 0.05$; **NS**); T7 / T8: $x2 = 7.01$ (*ddl* = 1; $P < 0.01$; **HS**); T7 / T9: $x2 = 2.78$ (*ddl* = 1; $P > 0.05$; **NS**); T8 / T9: $x2 = 0.32$ (*ddl* = 1; $P > 0.05$; **NS**).

Subplots treated with aqueous extract of *Carica papaya* **powder**: T6 / T7: $x2 = 6.86$ (*ddl* = 1; $P < 0.01$; **HS**); T6 / T8: $x2 = 2.56$ (*ddl* = 1; $P > 0.05$; **NS**); T6 / T9: $x2 = 3.86$ (*ddl* = 1; $P < 0.05$; **HS**); T7 / T8: $x2 = 2.56$ (*ddl* =

Treatments / Parameters	Witness	*C. papaya*	*S. didymobotrya*	*A. indica*	Optimal	*F (4,10)*
TF T6 (FL)	91,66±4,16 [Ba]	100±0,00 [a]	100±0,00[Aa]	100±0,00[Aa]	100±0,00[Aa]	4,00[*]
T7 (FID)	62,50±7,21[BCb]	75.00±0,00 [ABCb]	58,33±4,16 [Cb]	91,66±4,16 [A]	79,16±4,16[ABb]	8,58**
T8 (FvA)	100±0,00 [a]	100±0,00 [a]	100±0,00[a]	100±0,00[a]	97,43±2,56[a]	1,00[ns]
T9 (Fpsv)	100±0,00 [a]	71,42±14,28[b]	94,44±5,55[a]	94,44±5,55 [Aa]	100±0,00 [Aa]	2,64[ns]
F(3,8)	18,25**	4,72*	33,44***	1,44[ns]	16,92**	
NMG (%) T6 (FL)	154,31±13,29[a]	146,88±12,99[a]	119,56±10,66[a]	169,71±14,89[a]	147,58±13,14 [a]	0,01[ns]
T7 (FID)	72,73±6,44[ABb]	80,25±7,26[Ab]	45,26±4,03[Bb]	65,49±5,75[ABb]	41,19±3,63[ABb]	0,09[ns]
T8 (FvA)	222,84±20,64[a]	225,34±20,95[a]	182,90±16,76[a]	200,11±18,55 [a]	227,52±21,05 [a]	0,01[ns]
T9 (Fpsv)	90,66±7,76[b]	21,22±1,38[b]	67,65±5,96[ab]	89,46±8,00 [ab]	67,26±5,68 [b]	0,20[ns]
F(3,8)	0,25[ns]	0,46[ns]	0,33[ns]	0,24[ns]	0,42[ns]	
GN (%) T6 (FL)	66.88± 7.13 [a]	77,42±7,05 [a]	74,46±6,47 [a]	81,43±9,83 [a]	86,14±3,33 [a]	1,05[ns]
T7 (FID)	38,64±7,63[Bb]	55,11±5,78 [Ab]	34,83±2,6 [Bb]	54,041±3,98[Ab]	56,33±4,88 [Ab]	3,76*
T8 (FvA)	73,23±4,52 [a]	81,47±4,20 [ab]	87,73±5,45a	86,08±2,56a	83,27±1,36a	2,09[ns]
T9 (Fpsv)	63,74±5,59[a]	59,30±9,56[ABab]	68,64±5,03[a]	51,51±4,10[b]	75,44±9,90 [ab]	1,56[ns]
F(3,8)	5,71*	3,54 [ns]	19,55***	9,52**	5,34*	

1; $P > 0.05$; **NS**); T7 / T9: $x2 = 0.67$ (*ddl* = 1; $P > 0.05$; **NS**); T8 / T9: $x2 = 0.32$ (*ddl* = 1; $P > 0.05$; **NS**);

Subplots with aqueous extract of *Senna didymobotrya* **powder**: T6 / T7: $x2 = 8.08$ ($ddl = 1$; $P < 0.001$; **HS**); T6 / T8: $x2 = 2.24$ ($ddl = 1$; $P > 0.05$; **NS**); T6 / T9: $x2 = 1.23$ ($ddl = 1$; $P > 0.05$; **NS**); T7 / T8: $x2 = 1.24$ T8: $x2 = 1.24$ ($ddl = 1$; $P > 0.05$; **NS**); T7 / T9: $x2 = 0.19$ ($ddl = 1$; $P > 0.05$; **NS**); T8 / T9: $x2 = 0.06$ ($ddl = 1$; $P > 0.05$; **NS**) ;

Subplots treated with aqueous extract of *Azadirachta indica* **powder**: T6 / T7: $x2 = 2.09$ ($ddl = 1$; $P > 0.05$; **NS**); T6 / T8: $x2 = 7.11$ ($ddl = 1$; $P < 0.01$; **HS**); T6 / T9: $x2 = 101.94$ ($ddl = 1$; $P < 0.001$; **HRT**); T7 / T8: $x2 = 55.47$ ($ddl = 1$; $P < 0.001$; **HRT**); T7 / T9: $x2 = 17.33$ ($ddl = 1$; $P < 0.001$; **HRT**); T8 / T9: $x2 = 55.47$ ($ddl = 1$; $P < 0.001$; **HRT**);

Subplots treated with Optimal 20 sp: T6 / T7: $x2 = 5.58$ ($ddl = 1$; $P < 0.05$; **S**); T6 / T8: $x2 = 1.38$ ($ddl = 1$; $P > 0.05$; **NS**); T6 / T9: $x2 = -$ ($ddl = 1$; $P - $; $-$); T7 / T8: $x2 = 3.86$ ($ddl = 1$; $P > 0.05$; **S**); T7 / T9: $x2 = 4.70$ ($ddl = 1$; $P < 0.05$; **S**); T8 / T9: $x2 = 1.03$ ($ddl = 1$; $P > 0.05$; **NS**);

Comparison of the percentages of normal seeds in the different subplots :

Untreated subplots: T6 / T7: $x2 = 212.78$ ($ddl = 1$; $P < 0.001$; **HRT**); T6 / T8: $x2 = 68.03$ (*ddl = 1; P < 0.001;* *HRT)* $= 1$; $P < 0.001$; **HRT**); T6 / T9: $x2 = 290.03$ ($ddl = 1$; $P < 0.001$; **HRT**); T7 / T8: $x2 = 98.13$ ($ddl = 1$; $P < 0.001$; **HRT**); T7 / T9: $x2 = 12.24$ ($ddl = 1$; $P < 0.001$; **HRT**); T8 / T9: $x2 = 149.25$ ($ddl = 1$; $P < 0.001$; **HRT**).

Subplots treated with aqueous extract of *Carica papaya* **powder**: T6 / T7: $x2 = 10.44$ ($ddl = 1$; $P < 0.01$; **HS**); T6 / T8: $x2 = 34.18$ ($ddl = 1$; $P < 0.001$; **THS**); T6 / T9: $x2 = 1.27$ ($ddl = 1$; $P < 0.05$; **NS**); T7 / T8: $x2 = 16339.84$ ($ddl = 1$; $P < 0.001$; **HRT**); T7 / T9: $x2 = 6887.00$ ($ddl = 1$; $P < 0.001$; **HRT**); T8 / T9: $x2 = 31.61$ ($ddl = 1$; $P < 0.001$; **HRT**);

Subplots treated with aqueous extract of *Senna didymobotrya* **powder**: T6 / T7: $x2 = 60.25$ ($ddl = 1$; $P < 0.001$; **HRT**); T6 / T8: $x2 = 0.00$ ($ddl = 1$; $P > 0.05$; **NS**); T6 / T9 : $x2 = 99.26$ ($ddl = 1$; $P < 0.001$; **HRT**); T7 / T8: $x2 = 71.27$ ($ddl = 1$; $P < 0.001$; **HRT**); T7 / T9: $x2 = 1.02$ ($ddl = 1$; $P > 0.05$; **NS**); T8 / T9: $x2 = 123.18$ ($ddl = 1$; $P < 0.001$; **HRT**);

Subplots treated with aqueous extract of *Azadirachta indica* **powder**: T6 / T7: $x2 = 152.87$ ($ddl = 1$; $P < 0.001$; **HRT**); T6 / T8: $x2 = 196.93$ ($ddl = 1$; $P < 0.001$; **HRT**); T6 / T9: $x2 = 1898$ ($ddl = 1$; *P < 0.001;* $-$); T7/ T8: $x2 = 429.88$ ($ddl = 1$; $P < 0.001$; **HRT**); T7 / T9: $x2 = 71.85$ ($ddl = 1$; $P < 0.001$; **HRT**); T8 / T9: $x2 = 1296.14$ ($ddl = 1$; $P < 0.001$; **HRT**);

Subplots treated with Optimal 20 sp: T6 / T7 : $x2 = 318.62$ ($ddl = 1$; $P < 0.001$; **HRT**); T6 / T8 : $x2 = 1.40$ ($ddl = 1$; $P > 0.05$; **NS**); T6 / T9 : $x2 = 71.85$ ($ddl = 1$; $P < 0.001$; **HRT**); T7 / T8: $x2 = 429.88$ ($ddl = 1$; $P < 0.001$; **HRT**); T7 / T9: $x2 = 64.73$ ($ddl = 1$; $P < 0.001$; **HRT**); T8 / T9: $x2 = 2156.35$ ($ddl = 1$; $P < 0.001$; **HRT**).

Comparison of the average number of seeds per fruit in the different subplots :

Untreated subplots: T6 / T7: $t = 64.08$ ($ddl = 240$; $P < 0.001$; **HRT**); T6 / T8: $t = 34.61$ ($ddl = 314$; $P < 0.001$; **HRT**); T6 / T9: $t = 49.88$ ($ddl = 273$; $P < 0.001$; **HRT**); T7 / T8: $t = 81.12$ ($ddl = 225.85$; $P < 0.001$; **HRT**); T7 / T9: $t = 16.85$ ($ddl = 185$; $P < 0.001$; **HRT**); T8 / T9: $t = 71.72$ ($ddl = 258$; $P < 0.001$; **HRT**).

Subplots treated with *Carica papaya* **powder extract**: T6 / T7: $t = 51.07$ ($ddl = 241$; $P < 0.001$; **HRT**); T6 / T8: $t = 35.31$ ($ddl = 248$; $P < 0.001$; **HRT**); T6 / T9: $t = 118.71$ ($ddl = 224$; $P < 0.001$; **HRT**); T7 / T8: $t = 77.66$ ($ddl = 239$; $P < 0.001$; **HRT**); T7 / T9: $t = 74.06$ ($ddl = 157$; $P < 0.001$; **HRT**); T8 / T9: $t = 69.65$ ($ddl = 222$; $P < 0.001$; **HRT**).

Plots treated with *Senna didymobotrya* **powder extract**: T6 / T7: $t = 66.76$ ($ddl = 174$; $P < 0.001$; **THS**); T6 / T8: $t = 35.31$ ($ddl = 248$; $P < 0.001$; **THS**); T6 / T9: $t = 45.24$ ($ddl = 211$; $P < 0.001$; **HRT**); T7 / T8: $t = 84.31$ ($ddl = 168$; $P < 0.001$; **HRT**); T7 / T9: $t = 25.63$ ($ddl = 132$; $P < 0.001$; **HRT**); T8 / T9: $t = 69.59$ ($ddl = 205$; $P < 0.001$; **HRT**).

Plots treated with *Azadirachta indica* **powder extract**: T6 / T7: $t = 78.89$ ($ddl = 244$; $P < 0.001$; **HRT**); T6 / T8: $t = 15.57$ ($ddl = 311$; $P < 0.001$; **HRT**); T6 / T9: $t = 59.57$ ($ddl = 291$; $P < 0.001$; **HRT**); T7 / T8: $t = 77.12$ ($ddl = 202$; $P < 0.001$; **HRT**); T7 / T9: $t = 23.41$ ($ddl = 181$; $P < 0.001$; **HRT**); T8 / T9: $t = 62.68$ ($ddl = 249$; $P < 0.001$; **HRT**)

Plots treated with Optimal: T6 / T7: $t = 89.49$ ($ddl = 199$; $P < 0.001$; **HRT**); T6 / T8: $t = 39.95$ ($ddl = 309$; $P < 0.001$; **HRT**); T6 / T9: $t = 65.36$ ($ddl = 236$; $P < 0.001$; **HRT**); T7 / T8: $t = 98.20$ ($ddl = 177$; $P < 0.001$; **HRT**); T7 / T9: $t = 30.76$ ($ddl = 122$; $P < 0.001$; **HRT**); T8 / T9: $t = 88.09$ ($ddl = 232$; $P < 0.001$; **HRT**).

The table shows that for the sub-plots treated at :

➢ Witness :

- Overall, the difference between fruiting rates in treatments 6, 7, 8 and 9 is highly significant ($F = 18.25$; $ddl1 = 3$; $ddl2 = 8$; $P < 0.01$) and significant in the percentages of normal seeds ($F = 5.71$; $ddl1 = 3$; $ddl2 = 8$; $P < 0.05$) for these same treatments;

- The fruiting rate was 95.62% in treatment 6 and 84.87% in treatment 7; the difference between these two percentages is significant ($x2 = 5.71$; $ddl = 1$; $P < 0.05$);

- the percentage of normal seeds was 91.66% in treatment 6 and 62.50% in treatment 7. in treatment 7; the difference between these two percentages is significant ($x2 = 5.78$; $ddl = 1$; $P < 0.05$).

➢ *Carica papaya* :

- Overall, the difference between fruiting rates in treatments 6, 7, 8 and 9 is significant ($F = 4.72$; $ddl1 = 3$; $ddl2 = 8$; $P < 0.05$) and non-significant ($F = 3.54$; $ddl1 = 3$; $ddl2 = 8$; $P > 0.05$) between the percentages of normal seeds in these same treatments;

- The fruiting rate was 100% in treatment 6 and 75.00% in treatment 7; the difference between these two percentages is highly significant ($x2 = 168.43$; $ddl = 1$; $P < 0.001$);

- The percentage of normal seeds was 98.91% in treatment 6 and 92.29% in treatment 7, the difference between these two percentages is highly significant ($x2 = 6.86$; $ddl = 1$; $P < 0.01$).

➢ *Senna didymobotrya* :

- Overall, the differences between fruiting rates in treatments 6, 7, 8 and 9 are very highly significant ($F = 33.44$; $ddl1 = 3$; $ddl2 = 8$; $P < 0.001$) and very highly significant in the percentages of normal seeds ($F = 19.55$; $ddl1 = 3$; $ddl2 = 8$; $P < 0.001$);

- The fruiting rate was 100% in treatment 6 and 58.33% in treatment 7; the difference between these two percentages is highly significant ($x2 = 8.08$; $ddl = 1$; $P < 0.001$);

- The percentage of normal seeds was 74.46% in treatment 6 and 34.83% in treatment 7; the difference between these two percentages is highly significant ($x2 = 60.25$; *ddl* = 1; $P < 0.001$).

➢ *Azadirachta indica* :

- Overall, the difference between fruiting rates in treatments 6, 7, 8 and 9 is non-significant ($F = 1.44$; *ddl1* = 3; *ddl2* = 8; $P > 0.05$) and highly significant for the percentages of normal seeds ($F = 9.52$; *ddl1* = 3; *ddl2* = 8; $P < 0.01$);

- The fruiting rate was 100% in treatment 6 and 91.66% in treatment 7; the difference between these two percentages is non-significant $x2 = 2.09$ (*ddl* = 1; $P > 0.05$);

- The percentage of normal seeds was 81.43% in treatment 6 and 54.04% in treatment 7; the difference between these two percentages is highly significant ($x2 = 152.87$; *ddl* = 1; $P < 0.001$).

➢ Synthetic insecticide (Optimal 20 SP) :

- Overall, the difference between fruiting rates in treatments 6, 7, 8 and 9 is highly significant ($F = 16.92$; *ddl1* = 3; *ddl2* = 8; $P < 0.01$) and so is the percentage of normal seeds ($F = 5.34$; *ddl1* = 3; *ddl2* = 8; $P < 0.01$) in these same treatments;

- The fruiting rate was 100% in treatment 6 and 79.16% in treatment 7; the difference between these two percentages is highly significant ($x2 = 146.17$; *ddl* = 1; $P < 0.001$);

- The percentage of normal seeds was 98.77% in treatment 6 and 94.73% in treatment 7; the difference between these two percentages is significant ($x2 = 5.58$; *ddl* = 1; $P < 0.05$).

Regarding the comparison between the different insecticides we note :

➢ At the level of flowers left to pollinate freely

- Overall, the difference in fruiting rates between the insecticide treatments and the control is significant ($F = 4.00$; *ddl1* = 4; *ddl2* = 10; $P < 0.05$), while the percentage of normal seeds for these same treatments is insignificant ($F = 1.05$; *ddl1* = 4; *ddl2* = 10; $P > 0.05$);

- the fruiting rate was 100% in all treatments except the negative control whose fruiting rate was 91.66%; the difference between these percentages is significant and not significant for the percentage of normal seeds.

- However, the difference is highly significant between the insecticide treatments (botanical and chemical) and the control at the level of permanently closed flowers, and significant for the percentage of normal seeds.

For flowers labelled and protected and then intended for exclusive visit by *Apis mellifera* and for flowers labelled and protected and then intended for opening and closing without visit by insect or any other organism the difference is not significant for fruiting rates and percentages of normal seeds.

In general for all insecticide treatments, including the control, there was no significant difference ($P > 0.05$) between these different treatments.

The average number of kernels per fruit related to the activity of floricultural insects under the influence of insecticide treatments showed no significant difference between the different treatments (T6, T7, T8 and T9) as well as for the different insecticide treatments.

Table 15 shows the average values of the effects of the different insecticide treatments on the yield parameters of *S. melongena*.

Table 15: Seed weight in Kg and seed yield (Kg.ha^{-1}) of *Solanum melongena*.

Treatments	Grain yield (Kg.ha^{-1})	Grain weight (Kg)
Witness	0.42 ± 0.16 [b]	20.01 ± 1.99 [b]
Carica papaya	0.45 ± 0.05 [ab]	22.71 ± 2.63 [ab]
Senna didymobotrya	0.45 ± 0.02 [ab]	22.67 ± 1.40 [ab]
Azadirachta indica	0.57 ± 0.01 [a]	$28,17 \pm 0,70$ [a]
Optimal	$0,40 \pm 0,04$ [b]	$21,05 \pm 0,80$ [b]
$F_{(4, 15)}$	3.51 [*]	3.51 [*]

Each value represents the mean $\pm$ MSE. ns = non-significant difference ($P > 0.05$). (-) estimation of F-value is not possible due to equal variance. (*P < 0.05) significant difference

Figure 33 shows normal and abnormal seeds of *S. melongena*.

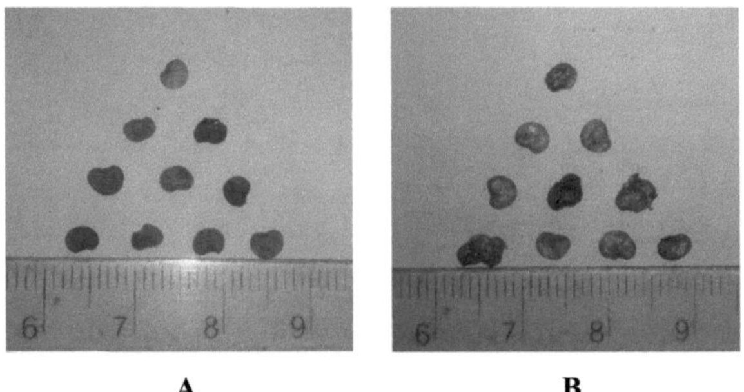

Figure 33: Normal (A) and abnormal (B) seeds of *Solanum melongena*

III.9. Cumulative impact of insecticide treatments and floricultural insects including *Apis mellifera* **on the yields of** *Solanum melongena*

The fruiting rates and percentages of normal seeds due to floricultural insects including *A. mellifera* in each treatment were 10.86% and 10.64% for *A. indica;* 5.33% and 3.01% for *C. papaya;* 6.48% and 6.48% for *S. didymobotrya*; 7.30% and 1.20% for Optimal and 4.88% and 5.09% for the control, respectively. The average number of seeds per fruit was 20.03%, 50.59%, 22.23%, 24.71% and 31.37% in the control and *C. papaya, S. didymobotrya, A. indica* and Optimal treated subplots respectively.

While collecting pollen from the flowers of this plant, the visitors were in contact with the anthers and stigmas. They carried the pollen from flower to flower on the same aubergine plant and on different plants. Thus, the significant increase in the fruiting rate and the percentage of normal seeds due to anthophilic insects, including *A. mellifera,* is the consequence of the activity of the latter on the pollination of this Solanaceae.

III.10. Pollination efficiency of *Apis mellifera* **on** *Solanum melongena* **yields**

The fruiting rates and percentages of normal seeds due to the influence of *A. mellifera* were 7.95% and 11.17% for *A. indica*; 4.06% and 2.27% for *C. papaya;* 4.24% and 4.88% for *S. didymobotrya*; 6.65% and 1.02% for Optimal and 6.35% and 7.87% for the control. The average number of seeds per fruit was 59.31%, 90.58%, 62.52%, 55.29% and 70.43% in the control and *C. papaya, S. didymobotrya, A. indica* and Optimal treated subplots respectively.

Thus, *A. mellifera* played a positive role in the fruiting and normal seed formation of this Solanaceae. The positive contribution of this bee to yields via its pollination efficiency would be justified by the action of the foragers during self-pollination and mainly during cross-pollination of the visited flowers. This proves that *A. mellifera* is one of the main insect pollinators of *S. melongena.*

CONCLUSION AND PERSPECTIVES

In sum, the present study is a contribution to the understanding of the impact of botanical insecticides (*Azadirachta indica, Carica papaya and Senna didymobotrya*) on pests and pollinators of *Solanum melongena* for better management in Cameroon. This work showed that *S. melongena* has an autogamous-allogamous mode of reproduction with a predominance of autogamy. In total, 15 insect species were found, including 12 species of insect pests, of which *Arhopalus rusticus was* the most frequent with 24.25% and preferentially attacks the leaves of this Solanaceae, and 3 insect species are pollinators, of which *A. mellifera was* the most frequent with 77.14% of visits.

Aqueous extracts of the leaf powders of *A. indica, C. papaya* and *S. didymobotrya* very significantly reduced the insect pest population of *Arhopalus rusticus*. The treated subplots allowed good foraging activity of pollinating insects.

The activity of *A. mellifera was* from 8 to 16 h with two peaks of activity located between 10 and 11 h for the negative and positive control sub-plots and the treated sub-plots except for those treated with *S. didymobotrya* whose peak is located between 12 and 13 h. On the flowers of this plant, the foragers of *A. mellifera* collected pollen exclusively. In general, the highest number of simultaneously active foragers is 1 per flower. The overall average abundance per 1000 flowers was 14 foragers, higher in the subplots treated with *C. papaya*. The overall average duration of an *A. mellifera* visit per flower is 8.01 sec in all subplots. The overall average foraging rate was 11 flowers per minute. The rate of visits by *A. mellifera* is positively correlated with the rate of flowering of *S. melongena*. This bee is faithful to the flowers of this Solanaceae during foraging trips.

The aqueous extract of *A. indica* is the most effective bio-insecticide with a yield of 0.57 Kg.ha^{-1} .

The data obtained make it possible to classify the aubergine as an exclusively pollinating bee plant. This Solanaceae can therefore be cultivated to increase pollen production as a hive product and to maintain bee colonies.

Comparing the yield of flowers left to pollinate freely with those protected from insects in the different insecticide treatments, there was an increase in fruiting rate, average number of seeds per fruit and percentage of normal seeds of 20.34%, 20.03% and 28.24% respectively for the control; 25%, 50.59% and 22.31% for *C. papaya;* 41.67%; 22.23% and 39.63% for *S. didymobotrya*; 8.344%; 24.71% and 27.39% for *A. indica* and finally 20.84%; 31.37% and 29.81% for Optimal.

For this plant, we showed that by their cumulative action, powdery mildew insecticides and floricultural insects increased the fruiting rate, the average number of seeds per fruit and the percentage of normal seeds which varied respectively from 10.86%; 24.71% and 10.64% for *A. indica; 5.33%; 50.59% and 3.01% for C. papaya indica;* 5.33%, 50.59% and 3.01% for *C. papaya;* 6.48%, 22.23% and 6.48% for *S. didymobotrya*; 7.30%, 31.37% and 1.20% for Optimal; and 4.88%, 20.03% and 5.09% for the control.

Through its positive action on the pollination of the visited flowers, the honey bee caused a significant increase in the fruiting rate, the average number of seeds per fruit and the percentage of normal seeds respectively of 0%, 59.31% and 12.95% for the control; 28.58%, 90.58% and 27.21% for *C. papaya*; 5.56%, 62.52% and 21.75% for *S. didymobotrya*; 5.56%, 55.29% and 40.16% for *A. indica* and finally 6.65%, 70.43% and 1.02% for Optimal. The average number of seeds per fruit was 59.31%, 90.58%, 62.52%, 55.29% and 70.43% in the control and *C. papaya, S. didymobotrya, A. indica* and Optimal treated subplots respectively.

In order to protect the aubergine crop from insect pests while preserving pollinating insects, the use of botanical insecticides should be encouraged as an alternative to the use of synthetic insecticides and to promote best production practices for this Solanaceae.

With this in mind, it would be wise to :

- to find the best concentration of these botanical insecticides to improve their efficacy;

- to test the efficacy of aqueous extracts of the leaves of *A. indica*, *C. papaya* and *S. didymobotrya* on other varieties of aubergine in other agro-ecological zones of Cameroon;

- to further study the pollination efficiency of *A. mellifera* on *S. melongena* under the influence of botanical insecticides.

- identify, isolate active molecules and study their mode of action.

BIBLIOGRAPHIC REFERENCES

A. Seck, 2019. Plant Resources of Tropical Africa, Ressources végétales de l'Afrique tropicale
Africa: p 1-8.

Abbé Coste, 2019. *Solanum melongena* L. eFlore, electronic flora of *Tela Botanica*, p 1-3.

Adamou M. & Tchuenguem F. F. - N., 2014. Foraging and pollination behavior of *Apis mellifera adansonii* Latreille (Hymenoptera, Apidae) on *Brachiaria brizantha* (Hochst. Ex A. Rich.) Stapf. 1919 flowers at Dang (Ngaoundere - Cameroon). *International Journal of Agronomy and Agricultural Research,* 4 (6) : p 62 - 74.

ADAMOU Moïse, 2017. Foraging and pollination activities of *Apis mellifera adansonii* on
flowers of *Bixa orellana* (Bixaceae) and *Brachiaria brizantha* (Poaceae) in Ngaoundéré
(Adamaoua, Cameroon). PhD thesis, Faculty of Science, University of
Ngaoundéré, p 148.

Adamou M., Yatahai C. M., Mazi S. & Tchuenguem F. - F. N., 2019. Pollination efficiency of
Apis mellifera (Hymenoptera: Apidae) on *Vigna unguiculata* (Fabaceae) flowers at Dang
(Ngaoundéré, Cameroon) at Djoumassi (North, Cameroon). *In: Book of Abstracts of the
26th Annual Conference of the Cameroon Bioscience Society*, Maroua, Cameroon, 26th -
30th Nov, p 222.

Agboyi, Tamo manuele & Ketoh Guillsume, 2018. Pesticide resistance in Plutella xylostella
(Lepidoptera: Plutellidae) populations from Togo and Benin. *Journal Tropical Insect
science,* 36 (4): pp 204 -210.

ALENE Chantal, Zéphirin TADU & Champlain DJIÉTO-LORDON 2019. Diversity of
hemipterans and ants in a market-gardening based agro-system in a suburb of Yaoundé,
Centre Region (Cameroon). *Int. J. Biol. Chem. Sci.* 13 (3): p 1669 - 1681.

Anjarwalla P, Belmain S, Sola P, Jamnadass R & Stevenson PC 2016. A guide to plants
Pesticides (Optimisation of pesticide plants: technology, innovation, awareness raising
&
networks) p 74.

Anonymous, 2002. Determination of sorption (adsorption-desorption) isotherms for
aubergine (solanum melongena l). *online brief,* p 2.

Anonymous, 2010. Soil fertility management for food security in Africa
Sub-Saharan Africa, p 45.

Anthony GINEZ & Thibault Verfaille, 2018. Integrated Biological Protection of Eggplant
under

Shelter, Series 2 Document2, p 9.

Arnaud AMELINE, 2019. Bottom-up approach to the regulation of aphid pests in orchards
Thesis UPV JAmiens, p 163.

Barry, Ngakou A. & Nukenine N. E. , 2017; Pesticidal Activity of Plant Extracts and a
Mycoinsecticide (*Metarhrizium anisopliae*) on Cowpea flower Thrips and Leaves
Damages
in the Field. *JEAI*, 18 (2) : 1 - 15, 2017; Article no. *JEAI*.37002.

Basga E., Fameni T. S. & Tchuenguem Fohouo F.-N., 2018. Foraging and pollination activity
of
Xylocopa olivacea (Hymenoptera: Apidae) on Vitellaria paradoxa (Sapotaceae) flowers
at
OuroGadji (Garoua, Camaroun), *Journal of Entomology and Zoology studies*. 6 (3):
p1015-
1022.

Basga, Sidonie FAMENI TOPE & Tchuenguem F. F - N, 2019. Pollination efficiency of *Apis*
mellifera Linné (Hymenoptera: Apidae) on the flowers of *Gossypium hirsutum*
(Malvaceae)
in Djamboutou (Garoua, Cameroon), p 14.

Binette & Jardin, 2019. Crop cards for fruit vegetables: aubergine, p 9.

BSV, 2019. Pests of vegetable crops p1.

Burkill H.M., 2000. The useful plants of west Tropical Africa. Royal Botanic Garden, Kiw, UK;
p 686.

Boni Barthélémy, Pierre Silvie & Françoise Komlan, 2017. Pesticide plants and the protection
of
vegetable crops in West Africa (bibliographic synthesis). *Biotechnol. Agron.*
Soc. Environ. 2017 **21** (4), p 288-300.

Charlotte Gourmel, 2014. Illustrated catalogue of the main insect pests and beneficials of
cultures   de   Guyane,   p   1-78.   .

Cheikh, EV coly & Mbaye diop, 2015. *Senna occidentalis* L., a promising plant in the fight
against
against *Caryedon serratus* Ol. (Coleoptera, Bruchidae), an insect pest of groundnut stocks
in Senegal. *International Journal of Biochemical and Chemical Sciences* 9 (3) : p 1399 .

CORAF, 2010. Plant extracts instead of synthetic insecticides. *Echoes of research*
regional N° 56, p 16.

DASSIE WENDJI ERIC, 2007. Effect of cropping practices on growth and productivity
of some market garden crops in the baïgom-foumbot area. Dissertation, FASA,
University of Dschang, p 117.

Daunay, M.C., Lester, R.N. & Ano, G., 2012. Cultivated aubergines. In: Charrier, A., Jacquot,
M., Hamon, S. & Nicolas, D. (Editors). Tropical plant breeding. Centre de
International Cooperation in Agricultural Research for Development (CIRAD) &
French Institute of Scientific Research for Development in Cooperation (ORSTOM),
Montpellier, France. p 83-107.

Denis DJONWANGWE, Joseph PANDO & F.-N. TCHUENGUEM FOHOUO, 2019.
Diversity
of floricultural insects of *Allium cepa* L. 1753 (Liliaceae) in Dogba and Gazawa (Far
North, Cameroon). *Journal of Chemical, Biological and Physical Sciences* 9 (2): p 281 -
295.

Denis-Esdras, Dodiomon SORO & Dossahoua TRAOR 2015. Evaluation of the infestation of
Loranthaceae on woody plants in agroecosystems in the Sud-Comoé region (Côte d'Ivoire)
d'Ivoire). *Int. J. Biol. Chem. Sci.* 9 (4): 1822-1834, p 13.

Demarly Y., 1977. Génétique et amélioration des plantes. Masson, Paris. 577 p.

Djakbé J. D., Ngakou A., Christian W., Faïbawe E., & Tchuenguem F. F - N., 2017.
Pollination
and yield components of *Physalis minima* (Solanaceae) as affected by the foraging activity
of
Apis mellifera (Hymenoptera: Apidae) and compost at Dang (Ngaoundéré, Cameroon).
International Journal of Agronomy and Agricultural Research, 11 (3): p 43 - 60.

Dounia & Tchuenguem F. F. -N., 2015. Foraging and pollination activities of *Apis mellifera,
Lipotriches collaris* and *Macronomia vulpina* (Hymenoptera: Apoidea) on flowers of
Glycine max (Fabaceae) and *Gossypium hirsutum* (Malvaceae) at Mayel - Ibbé (Maroua),
Cameroon). Doctoral thesis/PhD, University of Ngaoundéré, 148 p.

E.N. NUKENINE, Monglo B, Awasom I, Tchuenguem FFN & Ngassoum MB, 2002.
Farmers'
perception on some aspects of maize production, and some infestation levels of stored
maize
by *Sitophilus zeamais* in the Ngaoundéré region of Cameroon. *Cameroon Journal of*

Biological and Biochemical Sciences 12 (1), p 18-30.

E.N. NUKENINE, Adler C & Reichmuth Ch., 2010. Bioactivity of fenchone and *Plectranthus glandulosus* oil against *Prostephanus truncates* and two strains of *Sitophilus zeamais. Journal of Applied Entomology* 134 (20), p 132-141.

E.N. NUKENINE, Tofel H.K. & Adler C. , 2011. Comparative efficacy of neemAzal and local botanicals derived from *Azadirachta indica* and *Plectranthus glandulosus* against *Sitophilus zeamais* on maize. *Journal of pest science* 84 (4), p 479-486.

E.N. NUKENINE, Florence Pagou, Michael VABI & Cornel ADLER, 2013. Comparative of toxicity of four local botanical powders to *Sitophilus zeamais* and influence of drying regime and particle size on insecticidal efficacy. *Intertional Journal of Biological and Chemical Sciences* 7 (3), p 1313-1325.

FAOSTAT, 2016. FAO Statistics, Food and Agriculture Organization of the United Nations, Rome, 2020; p 37.

Fanou, Prudence Agnandji1, Boris Fresnel Cachon & Ménonvè Atindehou, 2018. Analysis of Phytosanitary practices in market gardening in intra-urban (Cotonou) and peri-urban areas (Sèmè-kpodji) in southern Benin. *African Journal of Environment and* Agriculture 2018 ; 1 (1), p 2 - 11.

François Meurgey, 2014. Diversity of wild bees in Guadeloupe and their contribution to the Florebutinea (Hymenoptera, Anthophila, Apidae and Megachilidae), p 4-65. .

Gabriel MAHBOU SOMO TOUKAM, 2010. Diversity of *Ralstonia solanacearum* in Cameroon Genetic basis of resistance in chilli (*capsicum annuum*) and Solanaceae. Thesis Institute of Life and Environmental Sciences and Industries, p 27.

GARUBA T., AZEEZ J. A., BELLO M. O. & LATEEF A. A, 2018. Isolation and physiological studies of fungi associated with post-harvest diseases of selected solanaceous fruits in ilorin markets, Nigeria. *Cameroon Journal of Biological and Biochemical Sciences* 2018, 26, 01-05.

HAL. Open Archives, 2016. Steroidal saponosides from aubergine (Solanum melongena L.) II. Variations in content related to harvesting conditions, genotypes and the amount of fruit seeds, p 59.

HOUADAKPODE DOSSA STANISLAS, 2018. Valorization of aromatic plants in the

   integrated management of the main insect pests of greater nightshade in southern benin: the

case of

   ocimum gratissimum and *o. Basilicum.* Dissertation, University of Liège, Gembloux, Agro-

Bio

   Tech, p 3-14.

IMFURA, 2019. How to grow aubergines like a pro, 3 p.

INRA, 2014. *Researchers fly to the rescue of bees.* Sauvignon, 28 p.

INRA, 2019. *Pesticides, agriculture and the environment*, p 68.

Isman MB. 2006. Botanical insecticides, deterrents and repellents in modern agriculture and an

   Increasingly regulated world. *Ann. rev. entomol*, 51: p 45-66.

Jangolo, 2017. All about eggplant-Solanum *melongena.* Blog, June 28 2017, p 3.

Jeanne Leborgne, 2019. Plants stronger than pesticides. *Plants and Health,* p 8.

Jérôme Lambion & GRAB, 2017. Biological control of mites and whiteflies on aubergine,

   Groupe de Recherche en Agriculture Biologique (GRAB), 2017; p 10.

Johnson Felicia, 2019. Impact of insects and nematodes on aubergine production

   Solanum aethiopicum Linné, 1756 Djamba F1 variety in the peri-urban area of Abidjan,

   Côte d'Ivoire. *International Journal of Multidisciplinary Research and* Development.

Volume

   6; January 2019; p 06-11.

Joseph Blaise Pando, Fernand-Nestor Tchuenguem & Denis Djonwangwe, 2018. Diversity

and

   impact of floricultural insects on fruit and grain yields of *Arachis hypogaea*

   L. 1753 (Fabaceae) in Maroua (Far North, Cameroon), p 13.

Joydeep, 2012. Description of the image of *Solanum melongena* 26-8-2012, p 36.

Kalengui, 2014. Determination of sorption (adsorption-desorption) isotherms for

   aubergine (solanum melongena l). *online brief,* p 2.

Kokopelli,2013. Aubergine technical data sheets. Aubergine flower and quantity of insects

   pollinator, p2.

L. Mwaura, P.C. Stevenson, D.A. Ofori, P. Anjarwalla, R. Jamnadass & P. Smith, 2013.

   Pesticidal plants Leaflet Tephrosia vogelii Hook. F, p 2.

Ladoh-Yemeda, Vandi T, Mpondo Mpondo E & Tomedi Eyango, 2016. Ethnobotanical

study.

of medicinal plants sold in the markets of the city of Douala, Cameroon.

Journal of Applied Biosciences 99: 9450 - 9466, p 17.

The Economy, 2021. *Journal du quotidien de l'econmie* N° 02194, p 5.

Leolong Azelie, 2013. Biological protection of aubergine: interest of relay plants in market gardening

biological. Master's thesis, University Digital Student Repository, p 28.

Lester & Seck, 2019. Plant resources of tropical Africa. PROTA, p 11.

Lucie Ayi-Fanou, 2018. Analysis of phytosanitary practices in market gardening in zones

intra-urban (Cotonou) and peri-urban (Sèmè-kpodji) in southern Benin. *African Review Environment and Agriculture* 2018; 1(1), p 2-11.

Marthe, 2016. Growing aubergine, chili and pepper successfully; p 1.

Mathilde Baude, 2011. Plants and pollinators observed in the wastelands of Seine-Saint- Michel Denis, p 4-35.

Mi-aime, 2019. *Solanum melongena* L, Aubergine or Bringelle, vegetable from La Réunion, Family :

Solanaceae - Solanaceae, p 2.

Mitali Das & Nilotpal Barua, 2013. Pharmacological activities of *Solanum melongena* Linn. (Brinjal plant). *International Journal of Green Pharmacy* 2013, p 274-277.

Mondédji A. D., 2015. Analysis of some aspects of the vegetable production system and perception of producers of the use of botanical extracts in the management of insect pests of vegetable crops in southern Togo. *Intenational Journal Biological Chemical Sciences*, 9 (1), p 98 - 107.

Mondédji A. D., 2015. Efficacy of extracts of neem leaves *Azadirachta indica* (Sapindale) on *Plutella xylostella* (Lepidoptera: Plutellidae), *Hellula undalis* (Lepidoptera: Pyralidae) and *Lipaphis erysimi* (Hymenoptera: Aphididae) of *Brassica oleraceae* (Brassicaceae). *International Journal of Biological and Chemical Sciences* 8 (5): p 2289.

Mpondo, Didier Siegfried DIBONG & Alfred NGOYE, 2012. Current state of medicine in the health system of the rural and urban populations of Douala (Cameroon). *Journal of Applied Biosciences* 55: 4036- 4045, p 10.

Muliele Tony, Constantine Manzenza & Aimé Mundele, 2017. Use and management of pesticides in market gardening: the case of the Nkolo area in the province of Central Kongo, Democratic Republic of Congo. *Journal of Applied Biosciences* 119: p 11954 - 11972.

Mwanauta R. W., Mtei K. M. & Ndakidemi P. A., 2015. Potential of Controlling Common Bean

Insect Pests (Bean Stem Maggot (*Ophiomyia phaseoli*), Ootheca (*Ootheca bennigseni*) and Aphids (*Aphis fabae*)) Using Agronomic, Biological and Botanical Practices in Field. *Agricultural Sciences*, 6: p 489-497.

Ngakou, Barry Borkeum Raoul & Nukenine Elias N. , 2019; 7(5): 333-338. The incidence of aqueous neem leaves (Azadirachta indica) extract and Metarhizium anisopliae Metch on cowpea thrips (Megolurothips sjostedti Trybom) and yield in Ngaoundéré (Adamaoua-Cameroon), p 7.

Ngamo TLS, 2004. In search of an alternative to the Persistent Organic Pollutants used for plant protection. *Bulletin d'informations phytosanitaires*. N° 43 (April-June), p 23.

Ngamo LS & Hance T. 2007. Commodity pest diversity and alternative control methods in a tropical environment. *Tropicultura*, 25: p 215-220.

OECD/FAO, 2019. OECD and FAO Agricultural Outlook 2019-2028, OECD Publishing, Paris/FAO, Rome, 352 p.

EPPO, 2019. A new insect pest of aubergine. *Association for the promotion of gardening, improving the environment and the living environment of plants and animals*, p 2.

Paris, 2016. Study on the action of aubergine pollinators in France, p 3.

Pando, Denis Djonwangwe, Jean Messi & F.N.Tchuenguem Fohouo, 2020. Diversity of floricultural insects of *Abelmoschus esculentus* (Malvaceae) and their impact on yields fruit and grain in Maroua Cameroon. *Journal of Animal & Plant Sciences (J.Anim.Plant Sci.* ISSN 2071-7024) Vol.43 (1): p 7350-7365.

Pierre Sylvie & Pierre martin, 2017. Pesticide plants to help crops. The CONVERSATION *Academic rigour, journalistic flair*, p 4.

Project, 2015. Biodiversity for agriculture in France, aubergine collection, p 6.

RADHORT, 2012. Integrated Production and Protection applied to vegetable crops in Sudano-Sahelian Africa, 158 p.

Son D., Somda I., Legreve A. & Schiffers B., 2017. Phytosanitary practices of tomato producers Burkina Faso tomato producers and risks to health and the environment. *Cahier Agriculture*, 26 (2): p 25005 - 25008.

Sreekanth, 2013. Field evaluation of certain leaf extracts for the control of mussel scale (*Lepidosaphes piperis* Gr.) in Black pepper (*Piper nigrum* L.). *Journal of Biopesticides*, 6 (1): p 1 - 5.

Stéphane Louabé, B. M. Kingha, B.S. Kengni & FFN Tchuenguem 2020. Foraging activity and pollination of *Papilio demodocus* (Lepidoptera: Papilionidae) on flowers of Lantana camara (Verbenaceae) in Dang (Ngaoundéré, Cameroon). *Journal of Chemical, Biological and Physical Sciences*, 2020, Vol. 10, No. 2; p 227-241.

Suzanne, 2009. PIP-Guide to good phytosanitary practices for aubergine cultivation PIP & Naeve, p 87.

Silvie Pierre, Yarou Boni Barthélémy & Francis François, 2017. Pesticide plants and protection of vegetable crops in West Africa (literature review) *Biotechnology, Agronomy, Society and Environment*, 21 (4): p 288 - 304.

Sylvia Salgon & Cyril Jourda, 2018. Genotyping by Sequencing Highlights a Polygenic Resistance to *Ralstonia pseudosolanacearum* in Eggplant (*Solanum melongena* L.). *int. J. Mol. Sci. , 19*, p 357 - 369.

TAIMANGA & Fernand-Nestor TCHUENGUEM FOHOUO 2018. Insect diversity floriculture and its impact on fruit and grain yields of *Glycine max* (Fabaceae) in Yassa (Douala, Cameroon). *Int. J. Biol. Chem. Sci.* 12 (1): p 141 - 156.

Tchuenguem F. F. - N., Messi J. & Pauly A., 2001. Activity of *Meliponula erythra* on flowers of *Dacryodes edulis* and its impact on fruiting. *Fruits*, 56: p 179 - 188.

Tchuenguem F. F. - N., 2005. Foraging and pollination activity of *Apis mellifera adansonii* Latreille (Hymenoptera: Apidae, Apinae) on the flowers of three plants in Ngaoundéré (Cameroon): *Callistemon rigidus* (Myrtaceae), *Syzygium guineense* var. *macrocarpum* (Myrtaceae) and *Voacanga africana* (Apocynaceae). Doctoral thesis, University of Yaoundé I, 103 p. .

TOFEL HAMAN Katamssadan, 2016. Insecticidal products from local Azadirachta indica A. Juss and Plectranthus glandulosus Hook for the protection of stored grains against the infestation of Callosobruchus maculatus F. and Sitophilus zeamais Motschulsky. Thesis of Doctorate / Ph. D, Faculty of Science, University of Ngaoundéré, 190 p.

Tsabang N, Nanga Ngah, Fokunang Tembe Estella & Agbor GA, 2016. Herbal Medicine and Treatment of Diabetes in Africa: Case Study in Cameroon. International Standard Serial Number : p 2572 - 5629 .

Williamson S., Ball A. & Pretty J., 2008. Trends in Pesticide Use and Drivers for Safer Pest Management in Four African Countries. *Crop Protection*, 27: p 1327 - 1334.

Yarou, Armel Mensah & Taofic Alabi 2017. Pesticide plants and crop protection in West Africa (bibliographic synthesis). *Biotechnol. Agron. Soc.*

Environ. 2017 **21** (4), p 288-304.

Yolou, Parisse AKOUANGO & Dominique ABOMO, 2018. Foundations and system of Market garden production in the commune of Athieme (south-west Benin). *Revue Espace, Territory, Society and Health;* Vol.3 - N° 5 June 2018, 35 p.

Zra Ganava Venceslas, Mazi Sanda & Tchuenguem Fohouo FernandNestor 2020. Pollination
efficiency of *Dactylurina staudingeri* (Hymenoptera: Apidae) on *Psorospermum febrifugum* (Hypericaceae) at Dang (Ngaoundéré, Cameroon). *Journal of Entomology and Zoology* Studies
2020; 8(1): p 216-224.

Appendix 1: Table of some pests and diseases of *Solanum melongena*

Type of Pests	Species	Family	Part of the plant plant attacked	Level of damage
	Agrotis spp.	Noctuidae	Leaves, fruits	+
	Thrips palmi OQ	Trpidae	Leaves, fruits	+++
	Epilachna spp.	Coccinellidae	Leaves, fruits	+++
	Epitrix cumumeris	Chrysomelidae	Sheets	++
Insects	*Bemisia tabaci OQ*		Leaves, fruits	+++
	Liriomyza trifolii OQ	Aleyrodidae Agromyzidae		++
	Aphis gossypii		Leaves, fruits	++
	Helicoverpa	Aphididae	Leaves, fruits	+++
	armigera OQ	Noctuidae		
	Amrasca spp.	Cicadellidae	Sheets	++
Mites	*Tetranychus spp.*			
Nematodes	*Meloidogyne spp.*	Tetranychidae		
		Heteroderidae		

Type of diseases	Species	Family	Part of the plant plant attacked	Level of damage
Stem and collar rot	*Pythium, Rhizoctonia*	Pyhtiacae/ Ceratobasidiaceae	Rods	+
Verticillium	*Verticillium dahliae*		Seeds	++
Alternaria	*Alternaria solani*	Plectospaerellaceae	Leaves, fruits, stems	+
		Pleosporaceae		
Late blight	*Phytophthora capsici*		Stems, foliage, fruits	+++
	P. parasitica	Peronosporaceae		
Powdery mildew	*Leveillula taurica,*		Sheets	+++
	Oidium longipes	Erysiphaceae		
Bacterial wilt	*Ralstonia*		Sheets	++
	solanacearum	Burkholderiaceae		

Appendix 2: Superfamily, family, scientific name, sponsor and year of description of the various insects cited in the brief

Superfamily	Family	Genus, species, subspecies, sponsor, year of description
Acridoidea	Acrididae	*Chorthippus brunneus* Thunberg, 1815
Aleyrodoidea	Aleyrodidae	*Trialeurodes vaporariorum* Westwood, 1856
Aphidoidea	Aphididae	*Aphis gossypii* Glover, 1877
Apoidea	Apidae	*Apis mellifera* Linnaeus, 1758
		Xylocopa caffra Linnaeus, 1767
		Xylocopa olivacea Fabricius, 1778
		Xylocopa inconstans Smith, 1874
		Xylocopa scioensi Gribodo, 1884
		Xylocopa gabonica Gribodo, 1894
	Halictidae	*Halictus sp.* Linnaeus, 1758
		Lasioglossum pullilabre Smith, 1874
	Megachilidae	*Chalicodoma cincta cincta* Fabricius, 1871
		Megachile rotundata Fabricius, 1787
		Megachile eurymera Smith, 1853
Asilidea	Asilidae	*Diogmites neoternatus* Bromley, 1951
Burkhoderiacae	Burkhoderiaceae	*Ralstonia solanacearum* Yabuuchi *et al.,* 1996
Cerambycoidae	Cerambycidae	*Arhopalus rusticus* Linnaeus, 1758
Cucujoidea	Coccinelledae	*Chelomenes sp.* Chevrolat, 1837
Chrysomelidea	Chrysomelidae	*Colorado beetle* Say, 1824
Coreoidea	Coreidae	*Colapsis sp.* Fabricius, 1801
		sp1 Leach, 1815
Cramboidea	Crambidae	*Acanthocephala femorata* Laporte, 1833
		Haritalodes derogata Fabricius, 1775
Dasypodaidea	Melittidae	*Hesperapis regularis* Cresson, 1878
	Nymphalidae	*Danaus plexippus* Linnaeus, 1758
		Hypolimnas misippus Ritsema, 1880
Mantoidea	Mantidae	*Mantis religiosa* Linnaeus, 1758
Meloidogygynidea	Meloidogygynidae	*Meloidogyne sp.* Goeldi, 1889
Membracoidea	Cicadellidae	*Euscelis sp.* Brullé, 1832
Noctuoidea	Noctuidae	*Helicoverpa armigera* Húbner, 1808
		Peridroma saucia Húbner, 1808
Pentatomoidea	Pentatomidae	(sp. 1) Leach, 1815
Pirrhocoroidea	Pirrhocoridae	*Dystercus sp.* Guerin-Meneville, 1831
Sarcophagoidea	Sarcophagoidae	*Sarcophaga carnaria* Linneaus, 1758
Sphingidea	Sphingidae	*Agrius sp.* Húbner, 1819
Syrphidea	Bombyliidae	*Bombylius major* Cresson, 1863
Tetranychidea	Tetranychidae	*Tetranychus urticae* Koch, 1836
Tettigonoidea	Tettigoniidae	*Tettigonia viridissima* Linnaeus, 1758
Vespoidea	Formicidae	*Crematogaster sp.* Lund, 1831

Appendix 3: Family, scientific name, sponsor and year of description of the various plants mentioned in the brief

Families	Scientific name, sponsor and year of description
Arecaceae	*Helianthus annus* Linnaeus, 1753 *Borassum aethiopium* Mart, 1838
Annonaceae	*Annona senegalensis* Persoon, 1836
Apocynaceae	*Voacanga africana* Stapf, 1857
Asteraceae	*Bidens pilosa* Linnaeus, 1753
	Tithonia diversifolia Gray, 1883
Burseraceae	*Boswelli dalzielü* Hutch, 1910 *Commiphora africana* Jacp, 1797
Caricaceae	*Carica papaya* Linnaeus, 1753
	Papaya edulis Linnaeus, 1753
	Papaya carica Gaertn, 1791
Cyperaceae	*Pennisetum glaucun* Pers, 1805
Commelinaceae	*Commelina diffusa* Burm, 1768
Fabaceae	*Cassia occidentalis* Link, 1829
	Phaseolus vulgaris Linnaeus, 1753
	Senna occidentalis Link, 1829
	Senna didymobotrya H.S. & Barberby,1982
	Trillogena anguina Lindley, 1836
	Vigna unguiculata Walp, 1843
	Medicago anguina Zarco, 1996
Lamiaceae	*Ocimum americanum* Linnaeus, 1755
	Ocimum basilicum Linnaeus, 1753
	Largestroemia speciosa Pers, 1806
Malvaceae	*Ocimum gratissumum* Linnaeus, 1753
	Abelmoschus esculentus Moench, 1794
	Bombax costatum Pelleg & Vuillet, 1914
	Gossypium hirsutum Linnaeus, 1763
	Hibiscus sabdariffa Linnaeus, 1753
Meliaceae	*Azadirachta indica* Juss, 1830
	Melia azedarach Juss,1830
Myrtaceae	*Psidium guajava* Linnaeus, 1753
Poaceae	*Andropognon goyanus* Kunth, 1833
	Cymbogon giganteus Sales, 2002 *Zea mays* Linnaeus, 1753
Solonacaea	*Solanum melongena* Linnaeus, 1753
	Solanum esculentum Dunal, 1813
	Solanum incanum auct. non L. Bock, 2019

Physalis minima Linnaeus, 1753

Appendix 4: Daily variation of the average temperature, hygrometry and number of visits of *Apis mellifera on Solanum melongena* flowers according to time slots (Bocklé, 2020).

Daily periods (hours)	6 - 7	8 - 9	10 - 11	12 - 13	14 - 15	16 - 17
Temperature (°C)	21,3	26,2	31,5	36,4	35	33,8
Humidity (%)	80,7	73,2	45,9	35,7	24,1	30
Number of visits	0	3	8	4	5	0

Appendix 5: Correlations between temperature, hygrometry and the number of visits of *Apis mellifera* (Bocklé, 2020).

Hours	6 - 7 h	8 - 9 h	10 - 11 h	12 - 13 h	14 - 15 h	16 - 17 h	Correlation between T (°C) and number of visits	Correlation between H (%) and number of visits
Temperature (°C)	21,3	26,2	31,5	36,4	35	33,8		
Humidity (%)	80,7	73,2	45,9	35,7	24,1	30		
Number of Control Visits	0	3	8	4	5	0	r = -0.03; DNS	r = -0.56; DTHS
Number of visits *Carica papaya*	1	4	9	3	0	0	r = -0.16; DNS	r = -0.62; DNS
Number of visits *Senna didymobotrya*	0	3	8	3	1	0	r =- 0.14; DNS	r = -0.55; DNS
Number of visits *Azadirachta indica*	0	5	13	14	2	0	r = 0.20; DNS	r = -0.41; DHS
Optimal number of visits	0	2	6	2	0	0	r = -0.20; DNS	r = -0.55; DHS

Appendix 6: Effect of insecticide treatments on the number of leaves per plant in the different treatments

Treatments	Days After Sowing (DAS)								$F_{(8,27)}$
	15	22	29	36	43	50	57	64	
Witness	$1,02 \pm 0,17^{Db}$	$1,78 \pm 0,02^{Dc}$	$3,17 \pm 0,24^{CDb}$	$6,10 \pm 0,49^{Cc}$	$7,24 \pm 0,24^{BCb}$	$10,53 \pm 0,62^{ABb}$	$12,60 \pm 1,69^{Ab}$	$13,30 \pm 1,56^{Ab}$	$22,52^{***}$
C. papaya	$1,16 \pm 0,8^{Db}$	$2,03 \pm 0,09^{Dc}$	$3,50 \pm 0,19^{Db}$	$7,56 \pm 0,59^{Cabc}$	$11,26 \pm 0,54^{Ba}$	$14,53 \pm 1,68^{ABab}$	$14,81 \pm 0,81^{Aab}$	$16,66 \pm 0,96^{Aab}$	$52,54^{***}$
S. didymobotrya	$1,38 \pm 0,00^{Fab}$	$1,38 \pm 0,00^{Fd}$	$3,94 \pm 0,13^{Eb}$	$6,39 \pm 0,51^{Dbc}$	$11,31 \pm 0,54^{Ca}$	$13,99 \pm 9066^{Bab}$	$14,99 \pm 0,86^{ABab}$	$15,47 \pm 0,62^{Aab}$	$134,26^{***}$
A. indica	$1,69 \pm 0,07^{Da}$	$2,54 \pm 0,47^{Db}$	$6,41 \pm 0,21^{CDa}$	$9,09 \pm 0,65^{BCa}$	$12,17 \pm 0,14^{Ba}$	$18,80 \pm 2,01^{Aa}$	$18,49 \pm 1,22^{Aa}$	$20,04 \pm 1,84^{Aa}$	$37,66^{***}$
Optimal	$1,34 \pm 0,18^{Dab}$	$2,95 \pm 0,15^{Da}$	$4,17 \pm 0,15^{Db}$	$8,70 \pm 0,60^{Dab}$	$10,32 \pm 0,56^{Ca}$	$16,82 \pm 0,89^{Cab}$	$17,34 \pm 1,09^{Ba}$	$18,73 \pm 1,77^{Aab}$	$86,98^{***}$
$F_{(4, 15)}$	$4,22^{*}$	$56,83^{***}$	$36,82^{*}$	$5,3^{**}$	$18,50^{***}$	$5,16^{***}$	$4,95^{***}$	$3,32^{**}$	

Each value represents the mean ± MSE. (* $P < 0.05$) significant difference; (** $P < 0.01$) highly significant difference; (*** $P < 0.001$) highly significant difference.

Appendix 7: Fruiting rate, average number of seeds per fruit and percentage of normal seeds of the different treatments of *Solanum melongena* (Bocklé, 2020).

		T6 (Free flowers)	T7 (Protected flowers)	T8 (FvA)	T9 (Fpsv)
Witness	NFE	24	24	40	20
	NFF	22	15	36	17
	mG/F	165,20	77,08	150,77	110,00
	NTG	3965	1850	6031	2173
	NGN	3096	1730	5100	2059
Carica papaya	NFE	24	24	40	20
	NFF	24	18	36	17
	mG/F	155,33	87,62	153,25	70,95
	NTG	3728	2125	6130	1419
	NGN	3388	1875	5763	1275
Senna didymobotrya	NFE	24	24	40	20
	NFF	23	15	33	17
	mG/F	127,83	48,37	121,95	85,50
	NTG	3068	1161	4878	1710
	NGN	2778	951	4418	1375
Azadirachta indica	NFE	24	24	40	20
	NFF	24	22	39	19
	mG/F	177,54	68,54	135,55	115,45
	NTG	4265	1645	5422	2309
	NGN	4172	1491	4934	1281

Optimal	NFE	24	24	40	20
	NFF	24	19	38	20
	mG/F	157,58	43,41	153,7	80,15
	NTG	3782	1042	6148	1603
	NGN	3466	737	5675	1344

More
Books!

OMNIScriptum

Printed by Books on Demand GmbH, Norderstedt / Germany